A Handbook of Mathematical Models with Python

Elevate your machine learning projects with NetworkX, PuLP, and linalg

Dr. Ranja Sarkar

‹packt›

BIRMINGHAM—MUMBAI

A Handbook of Mathematical Models with Python

Copyright © 2023 Packt Publishing

All rights reserved. No part of this book may be reproduced, stored in a retrieval system, or transmitted in any form or by any means, without the prior written permission of the publisher, except in the case of brief quotations embedded in critical articles or reviews.

Every effort has been made in the preparation of this book to ensure the accuracy of the information presented. However, the information contained in this book is sold without warranty, either express or implied. Neither the author, nor Packt Publishing or its dealers and distributors, will be held liable for any damages caused or alleged to have been caused directly or indirectly by this book.

Packt Publishing has endeavored to provide trademark information about all of the companies and products mentioned in this book by the appropriate use of capitals. However, Packt Publishing cannot guarantee the accuracy of this information.

Publishing Product Manager: Tejashwini R
Content Development Editor: Joseph Sunil
Technical Editor: Devanshi Ayare
Copy Editor: Safis Editing
Project Coordinator: Farheen Fathima
Proofreader: Safis Editing
Indexer: Rekha Nair
Production Designer: Jyoti Kadam
Marketing Coordinator: Vinishka Kalra

First published: August 2023

Production reference: 1180823

Published by Packt Publishing Ltd.
Grosvenor House
11 St Paul's Square
Birmingham
B3 1RB, UK.

ISBN 978-1-80461-670-3

www.packtpub.com

Dedicated to my Maa, Rina Sarkar, and my Baba, Nilambar Sarkar. I owe everything to my parents.

I would like to thank Souvik and Aarushi for being my support system.

– Ranja Sarkar

Contributors

About the author

Dr. Ranja Sarkar is a delivery manager (data science consulting) at TheMathCompany. She is a scientist with a focus on the intersection of data science and technology, working on a variety of projects in sectors ranging from manufacturing to healthcare to retail. Prior to this, Ranja worked as a data scientist at Shell plc, the most diversified global group of energy and petrochemical companies. Ranja has a decade of experience in academic scientific research encompassing physics, biophysics, statistical physics, and mathematics with publications in peer-reviewed internationally acclaimed journals. She has also worked at Deloitte offices in the US as a consultant for a year.

I want to thank my professors, mentors, colleagues, and friends for their continued support and encouragement.

To all the Packt editors, the managers who motivated me to write this content, and the entire team for keeping me on my toes throughout the process of writing the book and seeing it through to the end.

About the reviewer

Indraneel Chakraborty is a developer, specializing in creating data-driven workflows and high-quality web applications on cloud platforms. With a background in Bioinformatics and Biotechnology, he's skilled in academic research involving clinical trials registry data, and has published multiple research articles in reputed journals. In the biomedical software industry, he offers scientific and technical solutions while also developing highly scalable cloud infrastructure which includes automating and expediting large-scale processes. He is proficient in programming languages including Python and R, and has hands-on experience with Large Language Models (LLMs) and prompt engineering techniques. He is also an active contributor to open source projects and volunteers as a maintainer for open source courses that teach coding and data science skills to professionals and researchers worldwide. He also volunteers reviewing technical content. Indraneel's enthusiasm lies in exploring new technology stacks, honing his cloud engineering skills, and advancing his expertise in data science.

Table of Contents

Preface — xi

Part 1: Mathematical Modeling

1

Introduction to Mathematical Modeling — 3

Mathematical optimization	4	Formulation of the problem	8
Understanding the problem	4	Control theory	10
Formulation of the problem	5	Understanding the problem	11
Signal processing	7	Formulation of the problem	12
Understanding the problem	8	Summary	13

2

Machine Learning vis-à-vis Mathematical Modeling — 15

ML as mathematical optimization	16	Mathematical modeling – a prescriptive tool	24
Example 1 – regression	17	Finance	25
Example 2 – neural network	18	Retail	25
ML – a predictive tool	21	Energy	25
E-commerce	23	Digital advertising	26
Sales and marketing	24	Summary	26
Cybersecurity	24		

Part 2: Mathematical Tools

3

Principal Component Analysis — 29

Linear algebra for PCA	30	Applications of PCA	36
Covariance matrix – eigenvalues and eigenvectors	31	Noise reduction	38
		Anomaly detection	39
Number of PCs – how to select for a dataset	32	Summary	39
Feature extraction methods	34		
LDA – the difference from PCA	34		

4

Gradient Descent — 41

Gradient descent variants	43	Momentum	46
Application of gradient descent	43	Adagrad	47
Mini-batch gradient descent and stochastic gradient descent	45	RMSprop	48
		Adam	48
Gradient descent optimizers	46	Summary	49

5

Support Vector Machine — 51

Support vectors in SVM	52	Implementation of SVM	57
Kernels for SVM	54	Summary	60

6

Graph Theory — 61

Types of graphs	63	Weighted graphs	65
Undirected graphs	64	Optimization use case	66
Directed graphs	64		

Optimization problem	67	Graph neural networks	69
Optimized solution	67	Summary	71

Kalman Filter 73

Computation of measurements	75	Implementation of the Kalman filter	78
Filtration of measurements	76	Summary	82

Markov Chain 83

Discrete-time Markov chain	84	Gibbs sampling algorithm	86
Transition probability	84	Metropolis-Hastings algorithm	87
Application of the Markov chain	85	Illustration of MCMC algorithms	89
Markov Chain Monte Carlo	86	Summary	92

Part 3: Mathematical Optimization

Exploring Optimization Techniques 95

Optimizing machine learning models	96	Optimization in operations research	104
Random search	97	Evolutionary optimization	106
Grid search	100	Summary	107
Bayesian optimization	102		

Optimization Techniques for Machine Learning 109

General optimization algorithms	110	Second-order algorithms	113
First-order algorithms	112	Complex optimization algorithms	114

Differentiability of objective functions	114	**Summary**	**116**
Direct and stochastic algorithms	115	**Epilogue**	**116**

Index 119

Other Books You May Enjoy 124

Preface

Mathematical modeling is the art of transforming a business problem into a clear mathematical formulation. For a given problem or use case, the algorithmic implementation of a model helps optimize parameters and generate better insights and comprehension to enable decision-making. A mathematical model complements a machine learning model and supports high-stake decisions in sensitive domains such as medicine, for example.

There are three focal topics to help you understand mathematical modeling:

- Areas where a mathematical model is useful – for example, control engineering and signal processing
- Tested Python-based mathematical tools – for example, graph theory and MCMC
- Underlying algorithms of mathematical optimization

I will provide concepts of mathematical modeling and various approaches to modeling through this book. I will guide you in choosing the optimal technique and best-suited algorithm to solve a business problem using Python, based on two main sources of information:

- My experience from the past 5 years as a data scientist and application developer for businesses
- My academic research (at different stages of maturity) across science disciplines for a decade

As a data professional, I believe mathematical models (equation-driven) with objectives and constraints in a problem are as relevant as (data-driven) machine learning models. In some cases, the right combination of both yields the best solutions.

Who this book is for

Data scientists, research and development professionals, and business scientists, in general, can gain practical insights into mathematical modeling with Python from this book.

It is assumed that you have knowledge of the following:

- Differential equations
- Linear algebra
- Basics of statistics

- Data types and data structures
- Numerical algorithms

You will learn about the relevance of mathematical models, how interpretability must be factored into models while solving a business problem, and how mathematical optimization and tuning machine learning models are important to arrive at the optimal solution. You will also learn how to choose a model, keeping in mind the cost-effectiveness and efficiency of the underlying algorithm per the business case.

What this book covers

Chapter 1, *Introduction to Mathematical Modeling*, provides an introduction to the theory and concepts of mathematical modeling and the areas in which a mathematical model is predominant and useful.

Chapter 2, *Machine Learning vis-à-vis Mathematical Modeling*, describes with examples how machine learning models serve as predictive tools and classical mathematical models serve as prescriptive tools.

Chapter 3, *Principal Component Analysis*, provides the method to reduce the dimensionality of very high-dimensional data and examples wherein dimensionality reduction is necessary.

Chapter 4, *Gradient Descent*, is about an algorithm that lays the foundation for machine learning models. Variants of the gradient descent method are used to train machine learning as well as deep learning models.

Chapter 5, *Support Vector Machine*, provides mathematical details about an algorithm mostly utilized for data classification.

Chapter 6, *Graph Theory*, provides a theory that quantifies or models the relationships between interconnected entities in a network.

Chapter 7, *Kalman Filter*, is about a state estimation and prediction algorithm in the presence of imprecise and uncertain measurements of a dynamic system.

Chapter 8, *Markov Chain*, provides the theory of modeling a stochastic (random) process. The Markov chain is a class of probabilistic models that determines the next future state from knowledge of only the present state.

Chapter 9, *Exploring Optimization Techniques*, provides exposure to optimization algorithms used in machine learning models and those used in operations research. It also introduces you to evolutionary algorithms with examples.

Chapter 10, *Optimization Techniques for Machine Learning*, provides the methods for determining which algorithm to choose for the optimization of a machine learning model fitted to a dataset.

To get the most out of this book

You will need Python 3.0 or higher versions to run the code in respective chapters. Python libraries required to execute a particular method have been imported (compatible versions with Python 3.0) in the code, which can be installed readily in the notebook or Python environment of your system.

Software/hardware covered in the book	Operating system requirements
Python 3.0 or higher	Windows, macOS, or Linux
Python libraries	Windows, macOS, or Linux

If you are using the digital version of this book, we advise you to type the code yourself or access the code from the book's GitHub repository (a link is available in the next section). Doing so will help you avoid any potential errors related to the copying and pasting of code.

Download the example code files

You can download the example code files for this book from GitHub at `https://github.com/PacktPublishing/A-Handbook-of-Mathematical-Models-with-Python`. If there's an update to the code, it will be updated in the GitHub repository.

We also have other code bundles from our rich catalog of books and videos available at `https://github.com/PacktPublishing/`. Check them out!

Conventions used

There are a number of text conventions used throughout this book.

`Code in text`: Indicates code words in text, database table names, folder names, filenames, file extensions, pathnames, dummy URLs, user input, and Twitter handles. Here is an example: "The hyperparameter optimization methods in the `scikit-learn` Python library assume good performance scores are negative values close to zero."

A block of code is set as follows:

```
import pandas as pd, numpy as np
from collections import Counter
import matplotlib.pyplot as plt
```

When we wish to draw your attention to a particular part of a code block, the relevant lines or items are set in bold:

```
prediction = one_class_svm.predict(X_test)
prediction = [1 if i == -1 else 0 for i in prediction] #outliers denoted by 1, inliers by 0
print(classification_report(y_test, prediction))
```

> **Tips or important notes**
> Appear like this.

Get in touch

Feedback from our readers is always welcome.

General feedback: If you have questions about any aspect of this book, email us at customercare@packtpub.com and mention the book title in the subject of your message.

Errata: Although we have taken every care to ensure the accuracy of our content, mistakes do happen. If you have found a mistake in this book, we would be grateful if you would report this to us. Please visit www.packtpub.com/support/errata and fill in the form.

Piracy: If you come across any illegal copies of our works in any form on the internet, we would be grateful if you would provide us with the location address or website name. Please contact us at copyright@packtpub.com with a link to the material.

If you are interested in becoming an author: If there is a topic that you have expertise in and you are interested in either writing or contributing to a book, please visit authors.packtpub.com.

Preface   xv

Share Your Thoughts

Once you've read *A Handbook of Mathematical Models with Python*, we'd love to hear your thoughts! Scan the QR code below to go straight to the Amazon review page for this book and share your feedback.

```
https://packt.link/r/1-804-61670-2
```

Your review is important to us and the tech community and will help us make sure we're delivering excellent quality content.

Download a free PDF copy of this book

Thanks for purchasing this book!

Do you like to read on the go but are unable to carry your print books everywhere?

Is your eBook purchase not compatible with the device of your choice?

Don't worry, now with every Packt book you get a DRM-free PDF version of that book at no cost.

Read anywhere, any place, on any device. Search, copy, and paste code from your favorite technical books directly into your application.

The perks don't stop there, you can get exclusive access to discounts, newsletters, and great free content in your inbox daily

Follow these simple steps to get the benefits:

1. Scan the QR code or visit the link below

   https://packt.link/free-ebook/9781804616703

2. Submit your proof of purchase
3. That's it! We'll send your free PDF and other benefits to your email directly

Part 1: Mathematical Modeling

In this part, you will get to know the theory behind mathematical modeling. You will be introduced to the concepts of a mathematical model and how they are relevant in solving a business problem. A mathematical model relies heavily on domain knowledge, the objective of the business case formulated into a mathematical problem, and constraints in the context, while a machine learning (statistical) model relies on data. Mathematical modeling is complementary to machine learning; for some use cases, one is enough, whereas a few others need a blend of the two.

This part has the following chapters:

- *Chapter 1, Introduction to Mathematical Modeling*
- *Chapter 2, Machine Learning vis-à-vis Mathematical Modeling*

1
Introduction to Mathematical Modeling

There is a great deal of interesting work happening in data sciences, especially in the realms of **Machine Learning** (**ML**) and **Deep Learning** (**DL**), and they are popular for good reason. However, the more tried and tested old-timer, mathematical modeling, is not much talked about. Mathematical modeling methods are no less relevant and are complementary to ML. To create successful data products that solve real business problems, we must often deploy the whole breadth of available mathematical tools, far beyond ML.

A model is a simplified representation of a real system and captures the essence of the system. A mathematical model uses variables, operators, functions, equations, and equalities. Under the hood of mathematical models, there are first-principle models based on physical laws, stochastic models based on distributions, averages, and empirical models based on patterns or historical data. Based on the particular type of modeling, qualitative or quantitative recommendations can be made for the system under consideration. A mathematical model facilitates design and prototyping and substantiates decisions. To formulate a mathematical model, one needs the input and output, the constants and variables, the domain and boundary, or initial conditions and constraints. The solution can be analytic or numerical; in either case, it determines the typical behavior and critical parameters of the system, trends, dependency, and operating regimes. Systems can be deterministic, wherein we know the cause-effect relationship, or they may be stochastic, involving probability distributions.

A few mature tools in mathematical modeling are in the following areas:

- Mathematical optimization
- Signal processing
- Control theory

We will explore these mathematical modeling approaches in the following sections. A narrow focus on ML misses out on many relevant features of pure mathematical optimization in many use cases. Successful solutions across disparate domains blend the new world of ML with classical mathematical modeling techniques. For example, one can combine state-space modeling methods with ML to infer unobserved parameters of systems in a parameter estimation problem.

Mathematical optimization

A branch of applied mathematics is mathematical optimization, popularly known as mathematical programming. It finds applications in fields such as manufacturing, inventory control, scheduling, networks, economics, engineering, and financial portfolio allocation. Almost any classification, regression, or clustering problem can be cast as an optimization problem. Some problems are static, while some are dynamic, wherein the values of system variables change over time.

Understanding the problem

Mathematical optimization is basically choosing inputs from a set of allowed options to obtain the optimized or best possible output in a given problem. There are variables, which are essentially the decisions we have to make; constraints, which are the business rules we have to adhere to; and objectives, which are the business goals we are aiming to achieve by representing the real-world business problem as an optimization problem. For example, a hospital's business problem is equipment and facility capacity planning. Medical equipment including beds and testing kits comprise the decision variables in this case; constraints are conventional and crisis capacity levels and regulations; and finally, the objective is to maximize resource utilization and service performance and minimize operating costs at the same time.

The most basic optimization problem consists of an objective function or cost function, which is the output value we try to optimize, in other words, maximize or minimize. The inputs are variables that can be controlled. Variables can be either discrete or continuous. The scale of a problem is pretty much determined by the dimensionality, that is, the number of scalar variables (also called decision variables) on which the search is performed. Constraints or equations place limits on how big or small some variables can get. Some problems have constraints, which can be equality or inequality constraints, while some problems do not have them at all, which implies the unbounded optimization of the function.

A linear programming problem is an optimization problem wherein the objective function and all constraints are linear, that is, the variables have only first-order terms. It was linear programming that led to the development of optimization in the 1940s. If either the function or one or more constraint(s) is non-linear, then we have a non-linear programming problem. For example, optimizing smooth (well-defined gradient, continuous) functions is easier. Knowing the problem type enables the selection of the right tool to solve it.

Formulation of the problem

The general formulation of a mathematical problem with an objective function *f(x)* represents questions in terms of variables and constraints. A typical form is as follows:

Minimize f ($x_1, \ldots, x_n$) such that $f_i(x_1, \ldots, x_n) \leq 0$ where i = 1, 2, ..., m

The nature of variables and constraints can be quite diverse. The variables may be discrete, continuous, or sets (groups), and the constraints may be deterministic or stochastic. The objective function may also include dynamic aspects.

Sometimes we are interested in finding the global optimum point without any constraints or restrictions on the region in space. Such problems are unconstrained optimization problems. At other times, we have to solve problems subject to certain constraints, such as restrictions on control variables. For example, in the preceding case, we might have to minimize the function subject to ($x_1 + \ldots + x_n$) = 1 . These are constrained optimization problems.

Example 1:

Let us have multiple (inequalities) constraints with two variables, *x* and *y*, as follows:

2x + 3y ≤ 34

3x + 5y ≤ 54

0 ≤ x, y

A graphical optimization would be an overlap (dark region) of the graphs, shown in *Figure 1.1*. Here the constraints are linear, and therefore, the maximum and minimum must lie on the boundary. And it is most likely that the optimum solution occurs at one of the three specified points. With non-linear constraints, the optimum occurs either at the boundaries or between them. In unconstrained optimization, either the function has no boundaries, or if it has, they are soft.

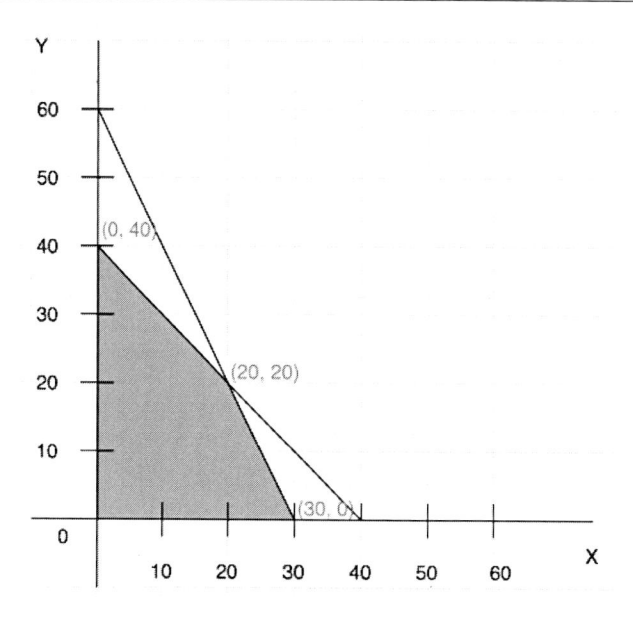

Figure 1.1: Graphical representation of linear constrained optimization

Typical constraints in business problems involve time, money, and resources while attempting to maximize an objective function. The constraints are more particular to the use case at hand while minimizing an objective function. Suppose in the preceding problem the objective function is linear, such as $f(x, y) = 20x + 35y$, and the optimum is found out from the slope of the function. If $f(x, y)$ takes a value, the value becomes a boundary, and the constraint plus the boundary make a linear constraint.

With linear constraints, the overlap region is considered to be feasible. Non-linear constraints can be very difficult to visualize as a distorted *x-y* plane makes it almost impossible to graph the feasible region.

Example 2:

In non-linear constrained optimization, the first step is to start on the boundary of the feasible region. To minimize the objective function, the vector direction should be chosen so that it decreases the function and stays in the feasible region. If the dot product of the gradient (slope) of the objective function with the vector itself is negative at a point on the boundary, then the vector is said to be moving in the descent direction. Also, a vector that does not violate the constraints is said to move in a feasible direction.

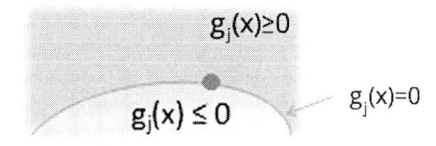

Figure 1.2: Feasible direction in non-linear constrained optimization

The constraint equation on the boundary is g(x) =0, shown in *Figure 1.2*. A feasible vector cannot cause the value of g(x) to increase. It must either remain zero or decrease. If the dot product of the gradient of the constraint with the vector itself is negative or zero at the point, then the vector is said to be moving in a feasible direction. For example, say we have the following objective function:

$$\text{Minimize} \quad f(x) = x_1 + 3x_2^2$$

And the initial point (4, 2√2) on a single constraint:

$$g_1(x) = x_1^2 + x_2^2 - 24 \leq 0$$

Where x_1 and x_2 are the variables, in general, standing for a matrix or array. The vector `<-1, 0>` is in both descent and feasible directions. Since the initial point is randomly chosen, there is a good chance that the overlap between the set of all feasible vectors and the set of all descent vectors is large. However, as we approach the minimum, the overlap gets smaller, and at the minimum or optimum point, there is no overlap at all. At the optimum, one cannot minimize the objective function further without violating the constraint. We know we have reached the optimum when the dot product of the two gradients is negative, and the two vectors have a matrix determinant equal to zero.

Another possibility is that the optimum occurs in the interior of the feasible region rather than on the boundary. In such a case, the gradient of the objective function will be zero at that point. The concavity (non-convexity) of the point is determined by the eigenvalues of Hessian (second differential) of the function.

In optimization problems where the objective function is noisy or its gradient is computed numerically as the gradient is not given (complex boundary value problems, for instance), errors are induced. Even if the objective functions themselves are not noisy, gradient-based optimization may turn out to be noisy. There are different optimizers available as library functions with Scientific Python, or `scipy` for short, to solve such optimization problems, and we will learn about a few of them in the following chapters. Now that we have learned about the concepts of mathematical optimization, we shall explore another concept in mathematical modeling, which is signal processing.

Signal processing

Another branch of applied mathematics is signal processing, which finds its application in the engineering field, focusing on analyzing and processing signals such as sound, images, scientific measurements, and filtering out noise. Signal processing deals with the transformation of a signal from time-series to hyper-spectral images, which are obtained from different electromagnetic measurements. Classic transformations of signals such as spectrograms and wavelets are often used with ML techniques. Such representations can also be used as inputs to deep neural networks. The Kalman filter is one classic signal processing filter that uses a series of measurements over time to produce estimates of unknown variables.

Understanding the problem

A signal is a function of a continuous variable, such as time or space. An analog signal is transformed into a digital signal by sampling it at specified intervals of time called the sampling period, the inverse of which is the sampling rate (per second or Hertz). The sampling rate has to be at least twice as high as the maximum frequency of the analog signal. It establishes a sufficient condition that permits a discrete sequence of samples to encapsulate all the information from a continuous time signal into a discrete time signal.

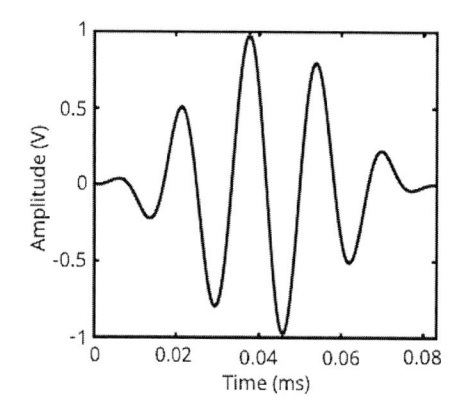

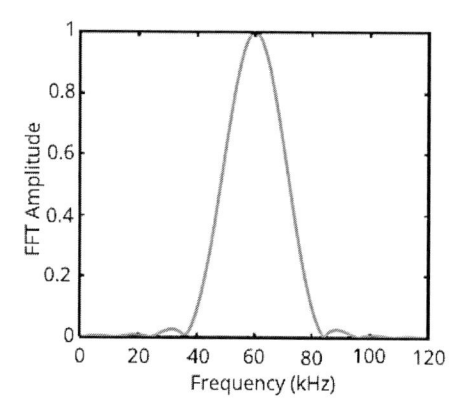

Figure 1.3: 60 kHz sinusoidal (Hann-windowed) tone burst in the time domain and frequency domain of the signal

The frequency domain representation of a signal is done with the **Discrete Fourier Transform (DFT)**. The **Fast Fourier Transform (FFT)** is an efficient computation method of DFT. FFT is rarely applied over the entire signal (speech signal, for example) at once but rather in frames due to the stochastic nature of the signal, an example of which is illustrated in *Figure 1.3*. FFT is available as a library function with `scipy` for the computation of the frequencies of each frame. A type of Fourier transform called the **Short-time Fourier Transform (STFT)** is typically applied on each individual frame.

Formulation of the problem

It is clear that **Discrete-Time Signal Processing (DSP)** is meant for sampled signals and establishes a mathematical basis for DSP, which is essentially analyzing and modifying a signal to improve (or optimize) its efficiency or performance. By using DFT, a discrete sequence can be represented as its equivalent frequency 'w' domain. The linearity property of the Fourier transform yields two signals, $x_1(t)$ and $x_2(t)$:

$$F[ax_1(t) + bx_2] = aX_1(w) + bX_2(w)$$

Where $X_1(w)$ and $X_2(w)$ are the Fourier transforms of $x_1(t)$ and $x_2(t)$ respectively, a concept often used in the filtering of signals, which is the transformation of the time t domain to the frequency w domain. The duality property of the Fourier transform is useful as it enables solving complex ones that otherwise would be difficult to compute directly. It yields that if x(t) has a Fourier transform $X(w)$, then one can form a new function of time $X(t)$ that has a functional form of the transformation, for example:

$$X(t) \leftarrow \rightarrow 2\pi x(-w)$$

A time shift affects the frequency, and a frequency shift affects the time of the functions. Let us take an example of a spectrogram to understand DSP.

A spectrogram displays the spectrum of frequencies of a waveform over time and is extensively used in the fields of music and speech processing and radars. It is generated by an optical spectrometer, a Fourier transform, or a wavelet transform and is usually depicted as a heat map wherein the strength or intensity of the signal changes with the color (brightness). To generate a spectrogram, a time-domain signal is divided into chunks of equal lengths that usually overlap, and FFT is applied to each chunk for the calculation of the frequency range. The spectrogram is a plot or graph of the spectrum on each segment or FFT frame, as a frequency *versus* a time image (or a 3D surface), shown in *Figure 1.4*, and the third dimension (represented by the color bar) indicates the amplitude of a particular frequency at a particular time. This process corresponds to the computation of the squared magnitude of STFT of the signal.

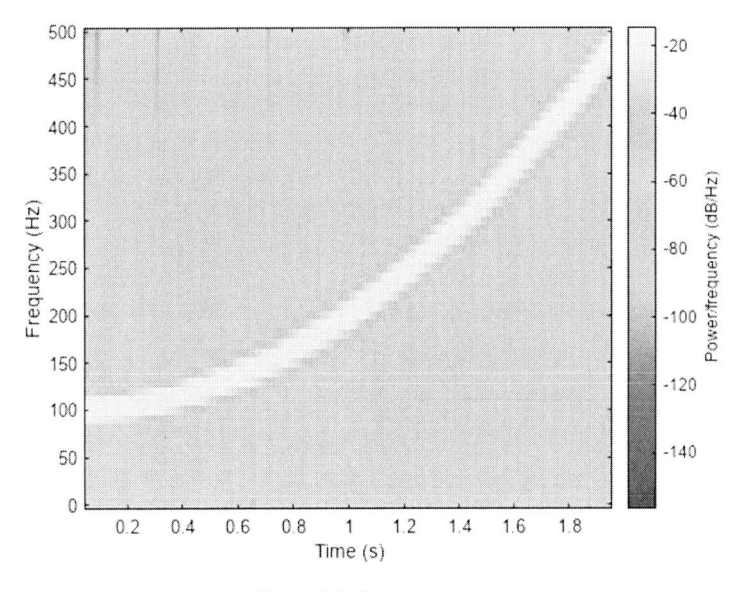

Figure 1.4: Spectrogram

Spectrograms can be used to identify characteristics of non-stationary or non-linear signals as a collection of time-frequency analyses. The parameters in a spectrogram typically are frame count (number of FFTs making it up), frequency range (minimum and maximum), FFT spacing, and FFT width (width of time each FFT represents).

Spectrograms are used with **recurrent neural networks** (**RNNs**) in speech recognition, as a primary example. We learned about how digital signals are free (well, almost) of noise and less distorted in this sub-section, and in the next, we are going to explore control theory, another mathematical modeling technique widely used in industrial processes. Control theory is, in general, useful whenever feedback happens in either regulator or servo mechanisms, for example, navigation systems and industrial production processes.

Control theory

A branch of mathematics and engineering is control theory, which found its use in social sciences as well, such as economics and psychology. It deals with the behavior or evolution of dynamical systems. It is particularly useful when the dynamics of a system are not arbitrary, that is, we understand the physics of the system. The objective of control is to develop a model from measured data. This model is a mathematical description of inputs applied to drive a system to a desired state, minimizing any delay or error simultaneously and ensuring a level of stability.

The behavior of a dynamical system is influenced by a feedback loop – a controller manipulates the system inputs to obtain the desired effect on the output. An error-controlled regulation is typically carried out with a **proportional-integral-derivative** (**PID**) controller, and as the name suggests, the signal is derived from a weighted sum, integral, and derivative of the error signal. The error, which is the difference between the actual and the desired output, is applied as feedback to the input. The standard terminology for a system is a process, and for a controlled variable is a **process variable** (**PV**), and the objective remains the reduction of the deviation error. Using a negative feedback loop, a measurement of PV (E in *Figure 1.5*) is deducted from a desired value S (set point or SP) to estimate an error (SP minus PV) in the system, which is used by a regulator R (*Figure 1.5*) to reduce the gap between the measured value and desired value. The error may be introduced into the system T as a disturbance D, as shown in the closed loop (*Figure 1.5*) of a controller.

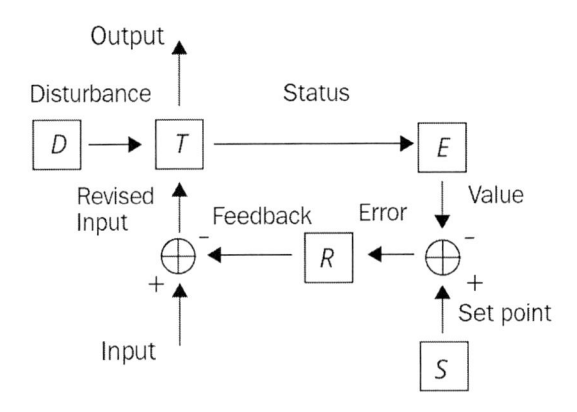

Figure 1.5: Negative feedback controller

Control theory can be linear as well as non-linear. Linear control theory is applied to devices obeying the superposition principle, meaning the output is roughly proportional to the input. Such (close to ideal) systems are tractable by frequency domain mathematical techniques such as Laplace transform, Fourier transform, and the Nyquist stability criterion. Non-linear control theory, on the other hand, applies to real-world systems that do not obey the superposition principle. Such systems are often governed by non-linear differential equations and analyzed using numerical methods. Non-linear systems are studied numerically using simulating operations using a simulation language that mirrors the system processes. However, if solutions in the vicinity of a stable or equilibrium point are only of interest, non-linear control systems can be linearized into approximations using perturbation techniques.

Understanding the problem

Mathematical techniques are served in either the frequency domain or time domain for analyzing control systems. The state variables in a frequency domain, representing the system's input, output, and feedback, are functions of frequency. The transfer function, system function, or network function is a mathematical model of the relationship between the input and output, on the basis of differential equations governing or describing the system. The input and transfer functions are converted from functions of time to functions of frequency by a mathematical transformation. In this domain, the differential equations are replaced by algebraic equations, which are simpler to solve. The state variables in a time domain are functions of time, and the system is described by one or more differential equations.

Time domain techniques are used to explore and analyze real non-linear systems because frequency domain techniques can only be used to study (ideal) linear systems. Although the equations for non-linear systems are difficult to solve, computer simulation methods have made their analyses commonplace. A critical application of the control loop is in industrial process control systems design, as shown in *Figure 1.6*.

Industrial process control loop

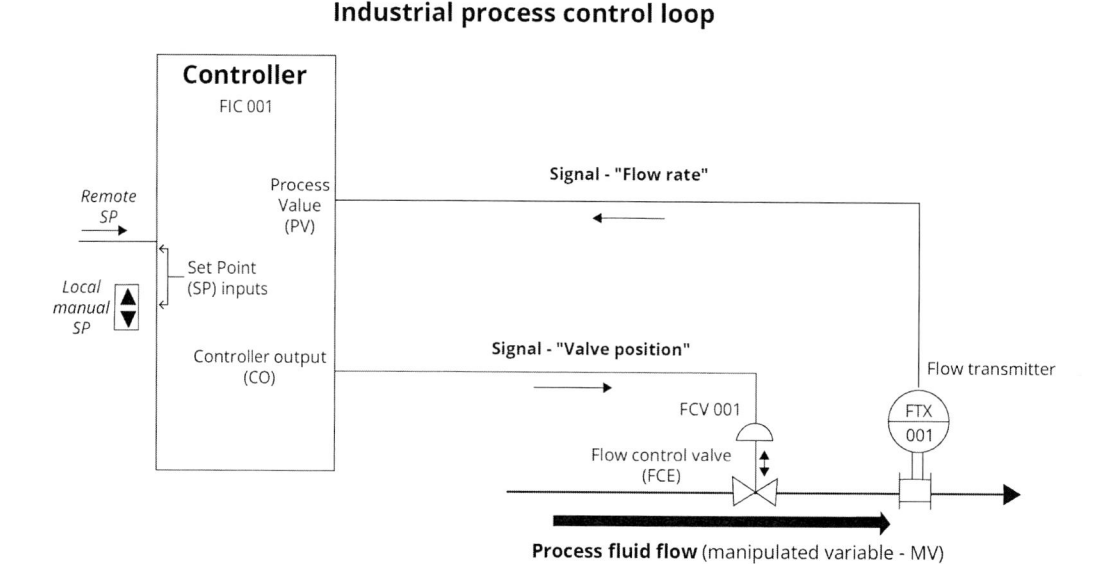

Figure 1.6: Industrial control showing continuously modulated process flow

The building block of industrial processes is the control loop, which consists of all elements to measure and control a process value at a desired SP in the presence of perturbances. The controller may be an isolated piece of hardware or, within a large distributed control system, a **programmable logic controller** (**PLC**) system and SP inputs can be manually set or cascaded from another source. The green text in *Figure 1.6* are tags that describe the function and identify a component and are unique (strings) within a plant representing the equipment components or elements. An associated sensor essentially captures the data of such tags.

Formulation of the problem

Modern control theory utilizes state-space methods (time-domain representation), unlike classical control theory, which uses transform methods (frequency-domain representation) such as the Laplace transform, which encodes all system information. In the state-space approach, a mathematical model is a set of first-order differential equations governing the related set of input, output, and state variables of the system. These variables are expressed as vectors, and the differential equations have a matrix format, which is more convenient to tackle. On the contrary, algebraic equations representing the behavior of a linear dynamical system are written in matrix form.

The state-space approach is not limited to linear systems and provides a convenient and compact way of modeling and analyzing mostly non-linear systems with multiple inputs and outputs. State space refers to a space whose axes are state variables, and the system state is expressed as a vector within that space.

A plant or process is the part of the system that is controlled, and the controller (or simply filter) makes up the rest. Inputs to the process have an effect on the outputs, and the effect is measured with sensors and processed by the controller. The control signal is fed back to the input, thus closing the loop. Such a typical architecture is the PID controller, which is by and large the most used industrial design, shown in *Figure 1.7*. It calculates an error value *e(t)* continuously, the error being the difference between the desired SP and measured PV, and applies a correction on the basis of proportional, integral, and derivative terms.

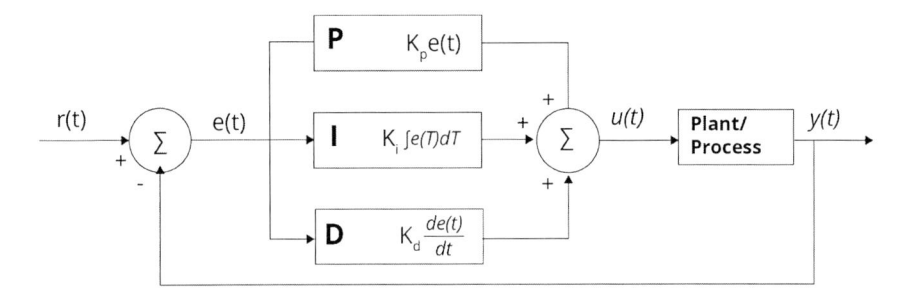

Figure 1.7: u(t) is the control signal sent to the system, e(t) = r(t) − y(t) is the error

When such a process is monitored by multiple controllers, it becomes a distributed control system with a decentralized control loop. Decentralization is useful as it helps the control systems to operate over a large area while interaction happens through communication channels.

Some of the main control techniques extensively used in industries include adaptive control, hierarchical control, optimal control, robust control, and stochastic control. Apart from these, intelligent control uses **artificial intelligence** (**AI**) and ML approaches such as fuzzy logic, neural networks, and so on to control a dynamic system. Industry 4.0 is revolutionizing the way manufacturers are integrating AI into their operations and production facilities.

Summary

In this chapter, we introduced the concepts of mathematical modeling via the important areas it is largely implemented in or applied to, such as optimization, signal processing, control systems, and control engineering. Mathematical modeling or mathematical programming is the art of transforming a problem into a clear mathematical formulation. Its subsequent algorithmic implementation generates actionable insights and helps build further knowledge about the domain.

The chapter helped us learn the formulation of a mathematical optimization problem in order to arrive at an optimal solution, the formulation being dependent on the domain we intend to investigate. A mathematical optimization model is like a digital twin of a real-world business scenario. It mirrors the business landscape in a strictly mathematical and programming setup, and such an environment becomes particularly relevant for the interpretability of business processes to support high-stake decisions.

In the next chapter, we will find out how mathematical models emphasize the importance of both data and domain knowledge. Additionally, we will learn how ML models can be cast as optimization problems.

2
Machine Learning vis-à-vis Mathematical Modeling

Having learned about the main components of mathematical optimization, which are decision variables, objective functions, and constraints, in the previous chapter, it is time to throw light on **machine learning** (**ML**) models, most of which can be cast as mathematical models. Humans make machines learn from huge amounts of historical data. ML models enhance the decision-making abilities of man and machine, exploiting the power of data and algorithms. There is almost always some optimization algorithm working in the background of most of these models.

The term ML was first popularized by Arthur L. Samuel in the 1950s, who was a pioneer in computer science and gaming. Data volume has increased by leaps and bounds since then, particularly in the last couple of decades, and making sense of huge amounts of data is beyond the scope of the human mind. Hence, ML stepped in and found its application in almost all domains to assist humans with the decision-making process.

Learning problems in data science can be broadly classified into regression, classification, and clustering depending on the business problem or use case. Regression and classification use supervised algorithms to predict a target, usually called the dependent variable, the independent variables being called predictors. Clustering makes use of unsupervised learning algorithms where the target is unknown. It is worth mentioning that learning in all ML algorithms is not all about optimization, an example of which is supervised learning in **k-nearest neighbors** (**kNN**). ML is a predominantly predictive tool helping a business plan for the future, thereby being beneficial for its bottom line. Businesses also leverage ML in anomaly (or outlier) detection and recommendation systems. Strictly mathematical modeling, on the other hand, helps businesses make decisions in areas such as electricity distribution, employee scheduling, and inventory management.

Some well-known algorithms used in ML models that employ constrained optimization are as follows:

- **Principal component analysis (PCA)**
- Clustering with an expectation maximization algorithm (a Gaussian mixture model, for example)
- Support vector machines using the method of Lagrange multipliers

Other ML algorithms that employ unconstrained optimization are **stochastic gradient descent (SGD)** in neural networks and batch gradient descent in deep learning (neural networks with numerous hidden layers between the input and output). Apart from these, there are genetic algorithms in evolutionary learning, which encompass both constrained and unconstrained optimization problems.

The main components of ML are representation, evaluation, and optimization. By representation, we essentially mean putting forth the knowledge and historical data statistically to find patterns, in other words, the formulation of a business problem to arrive at or estimate the solution. Next is the evaluation of the formulation, which we call the model, and fitting our data into and comparing it with known examples or data samples. Finally, the algorithm behind the model optimizes its weights and biases for a better fit with the data, and the optimization process iterates until a desired accuracy for the problem is attained. We will learn about PCA and gradient descent in the following chapters.

This chapter covers the following topics:

- ML as a mathematical optimization problem
- ML as a predictive tool
- Mathematical modeling as a prescriptive tool

ML as mathematical optimization

ML can be described as finding the unknown underlying (approximate) function that maps input examples to output examples. This is where the ML algorithm defines a parametrized mapping function and optimizes or minimizes the error in the function to find the values of its parameters. ML is function approximation along with function optimization. The function parameters are also called model coefficients. Each time we fit a model to a training dataset, we solve an optimization problem.

Each ML algorithm makes different assumptions about the form of the mapping function, which in turn influences the type of optimization to be performed. ML is a function approximation method to optimally fit input data. It is particularly challenging when the data (the size or the number of examples) is limited. An ML algorithm must be chosen in a way that it most efficiently solves an optimization problem; for example, SGD is used for neural nets, while ordinary least squares and gradient descent are used for linear regression. When we deviate from the default algorithms, we need a good reason to do so. In mathematical optimization, a heuristic might sometimes be used to determine near-optimal solutions. This happens when the classical algorithms are too slow to even find an approximate solution or they fail to find an exact solution to the optimization problem. Examples of heuristics are a genetic algorithm and a simulated annealing algorithm.

Example 1 – regression

An ML problem is framed as the learning of a mapping function (f) given input data (X) and output data (Y) such that Y = f(X). Given new input data, we should be able to map each datum onto (or predict) the output with our learned function, f. A prediction error is expected in general with noise in observed data and with a choice of learning algorithm that approximates the mapping function. Finding the set of inputs that results in the minimum error, cost, or loss is essentially solving the optimization problem. The choice of mapping function dictates the level of difficulty of optimization. The more biased or constrained the choice, the easier it is.

For example, linear regression is a constrained model. Using linear algebra, it can be solved analytically. The inputs to the mapping function in this case are the model coefficients. An optimization algorithm such as iterative local search can be used numerically but it is almost always less efficient than an analytical solution. A logistic regression (for a classification task) is a less constrained model, and an optimization algorithm is required in this case. The loss or error here is also called the logistic loss or cross-entropy. While a global search optimization algorithm can be used in both types of regression models, it is mostly less efficient than using either an analytical method or a local search method. An iterative global search (gradient descent, for example) is suitable when the search space or landscape is multimodal and nonlinear, as shown in *Figure 2.1*.

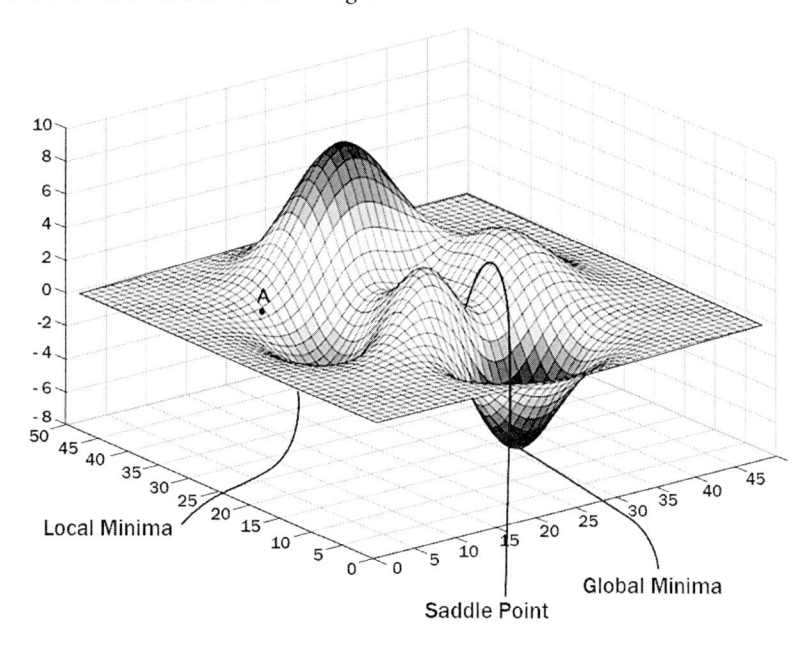

Figure 2.1: 3D landscape of unconstrained optimization space, where A is the current state

Example 2 – neural network

A neural network is a flexible model and imposes very few constraints. A network typically has an input layer, a hidden layer (can be more than one), and an output layer of nodes, and the inputs to the mapping function are weighted to the input layer, as shown in *Figure 2.2*. It is this mapping function that the supervised learning algorithm tries to best approximate.

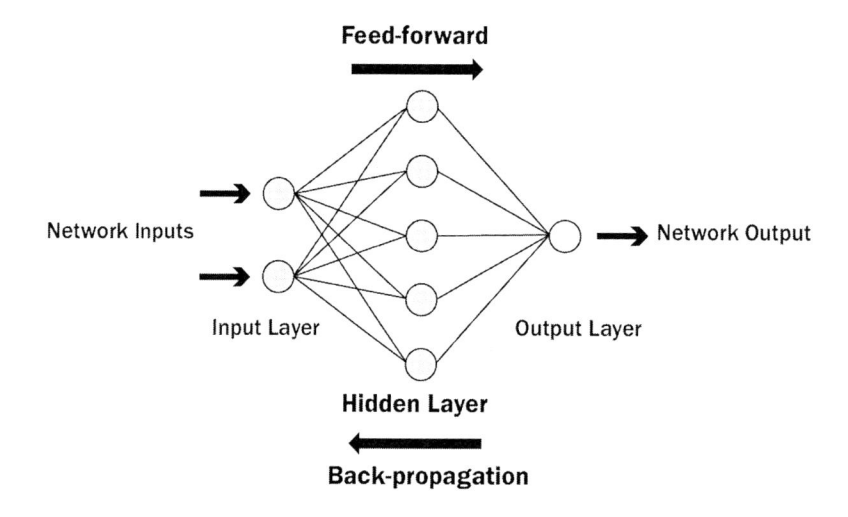

Figure 2.2: The three essential, minimal layers in a network

The deviation of predicted output from expected output is the error value, and this error or cost, shown in *Figure 2.3*, is minimized while approximating the function during model training. A neural network requires an iterative global search algorithm. Gradient descent is the preferred method to optimize a neural network that has variants, namely, batch and mini-batch gradient descent and SGD. One of the most popular SGD algorithms is **Adaptive Moment Estimation** (**Adam**), which computes adaptive learning rates for each parameter of the function.

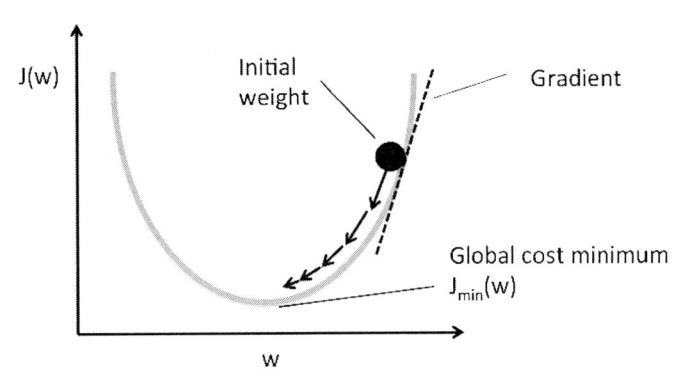

Figure 2.3: Minimization of cost function J(w) by gradient descent where w is the input (courtesy of Python Machine Learning by Sebastian Raschka)

A gradient is a vector of partial derivatives (slope/curvature) of the function with respect to input variable values. The gradient descent algorithm, as the name suggests, requires the calculation of this gradient. The negative of the gradient of each input is followed downhill as the gradient points uphill, to lead to new values of the input. A step size is used to scale the gradient and control the change of input with respect to the gradient. This step size or increment is the learning rate, a hyper-parameter of the algorithm, and is the proportion in which network weights are updated. The process is repeated until the minimum of the function is located. Gradient descent is adapted to minimize the loss function of a predictive model, such as regression or classification. This adaptation results in SGD, as shown in *Figure 2.4*.

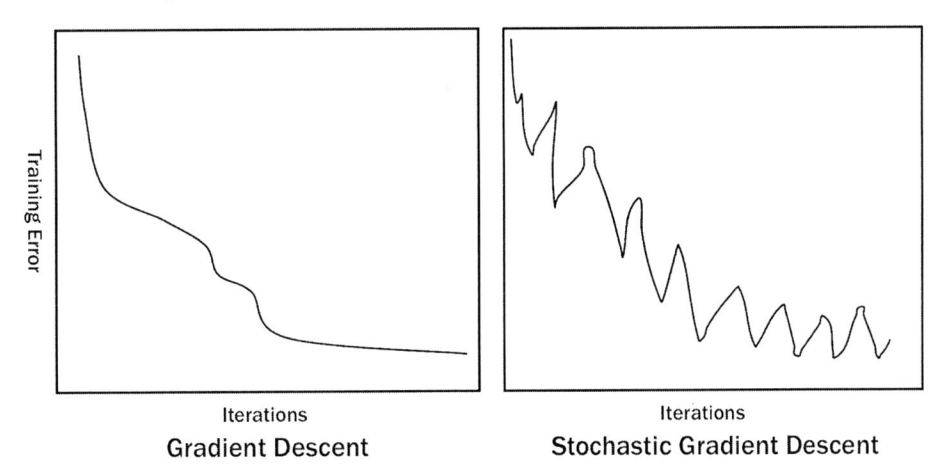

Figure 2.4: Gradient descent extension

SGD is the extension of the gradient descent optimization algorithm, wherein the target function is considered to be the loss or error, such as mean squared error for regression and cross-entropy for classification. Since the gradients of the target function with respect to the inputs are noisy, and deterministic to the extent of probabilistic approximation only, the algorithm is referred to as "stochastic." Due to the sparseness and noise in training data, the evaluated gradients have statistical noise. Generally speaking, SGD and its variants are still the most used optimization algorithms for ML as well as training deep learning (artificial neural network) models. The inputs to a neural network are the weights (model parameters) and the target function is the prediction error averaged over one batch, which is a subset of the training dataset.

A popular extension to SGD for the improvement of process efficiency, such as finding out the same (or better) loss in fewer iterations, is Adam. The Adam optimization method is computationally efficient, requires little memory, and is well suited for problems that are large in terms of size and features. The configuration parameters of Adam are the learning rate (step size), exponential decay rate (denoted by beta 1) for the mean (first moment) estimates, exponential decay rate (denoted by beta 2) for variance (second moment) estimates, and epsilon (very small number) to prevent any division by zero in the implementation. Larger values of learning rate (denoted by alpha) result in faster initial learning before an update and lower values of learning rate mean slower learning during the entire training. These parameters typically require very little tuning as they have intuitive interpretation.

A major challenge in using SGD to train a multi-layer neural network is the gradient calculation for nodes in the hidden layer(s) of the network. It can be tackled by utilizing a specific technique from calculus called the chain rule, and an efficient algorithm that implements this rule is called backpropagation, which calculates the gradient of a loss function concerning the model variables. The first-order derivative of a function for a specific input variable value is the rate of change of the function with that variable, and when there are multiple input variables, the (partial) derivatives form a vector. For each weight in the network, backpropagation calculates the gradient, which is then used by the SGD optimization algorithm to update the weights. Backpropagation works backward from the output toward the input of the network, as shown in *Figure 2.5*. It propagates the error in the predicted output to compute the gradient for each input variable, basically a backward flow of information from the cost function through the network. Backpropagation involves the recursive application of the chain rule, which is the calculation of the derivative of a sub-function given the known derivative of the parent function.

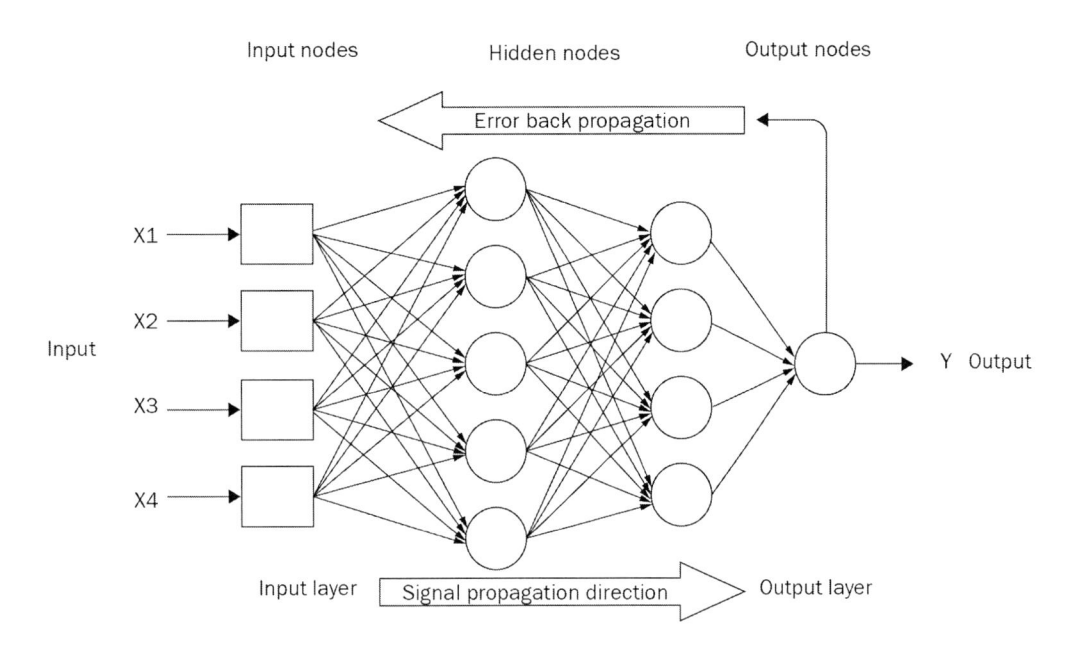

Figure 2.5: Backpropagation in a neural network

A genetic algorithm does not utilize the structure of the model, meaning it does not require gradients. For problems in which we use neural network models, we need to optimize the model using gradients that are calculated with backpropagation. It is only fair to say that backpropagation is a part of the optimization process, the optimization algorithm being SGD.

Now that we have explored ML tasks such as regression, classification, and neural nets in the form of mathematical optimization problems, we shall learn about ML as a predictive modeling tool and how it is utilized in a few important domains.

ML – a predictive tool

Working through a predictive model involves optimization at multiple steps on top of optimally fitting the learning algorithm to the data. It involves transforming raw data into a form most appropriate for consumption in learning algorithms. An ML model has hyperparameters that can be configured to tailor it to a specific dataset. It is a standard practice to test a suite of hyper-parameters for a chosen ML algorithm, which is called hyper-parameter tuning or optimization. A grid search or random search algorithm is used for such tuning. *Figure 2.6* shows the two search algorithm types. Grid search is more suitable for a quick search of hyperparameters and is known to perform well in general. You can also use Bayesian optimization for hyper-parameter tuning in some problems. We will learn about these optimization techniques in detail in the last part of the book.

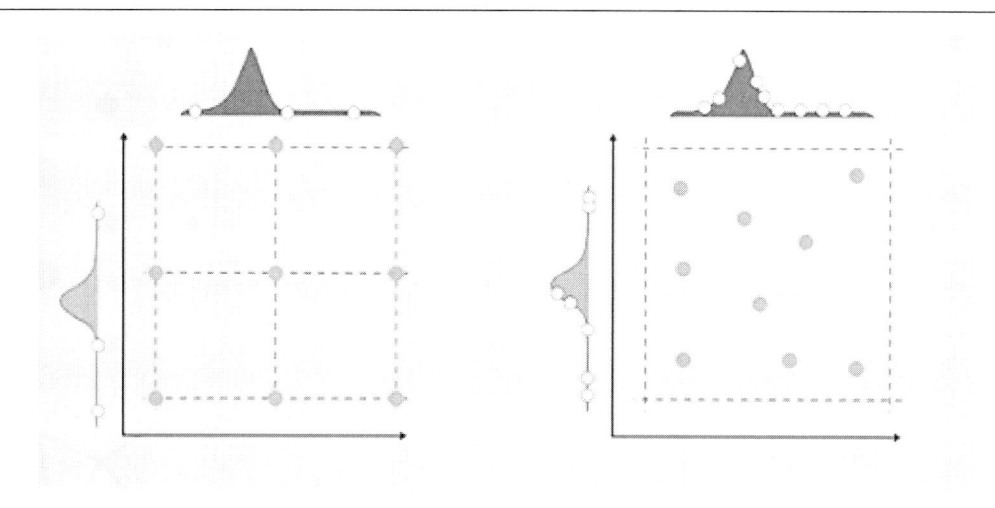

Figure 2.6: Grid search (L) versus random search (R)

An ML practitioner often performs a manual process for predictive model selection involving tasks such as data preparation, evaluating models, tuning them, and finally, choosing the best model for a given dataset. This can be framed as an optimization problem that can be solved with **automated machine learning** (**AutoML**) with little user intervention. The automated optimization approach to ML is also offered as a cloud product service by companies such as Google and Microsoft.

With or without a target variable in the input dataset, an ML algorithm becomes supervised or unsupervised learning, respectively. In reinforcement learning, certain behaviors are encouraged and others discouraged. The desired behavior is reinforced by rewards, which are gained through experiences from the environment. These three types of ML are shown in *Figure 2.7*.

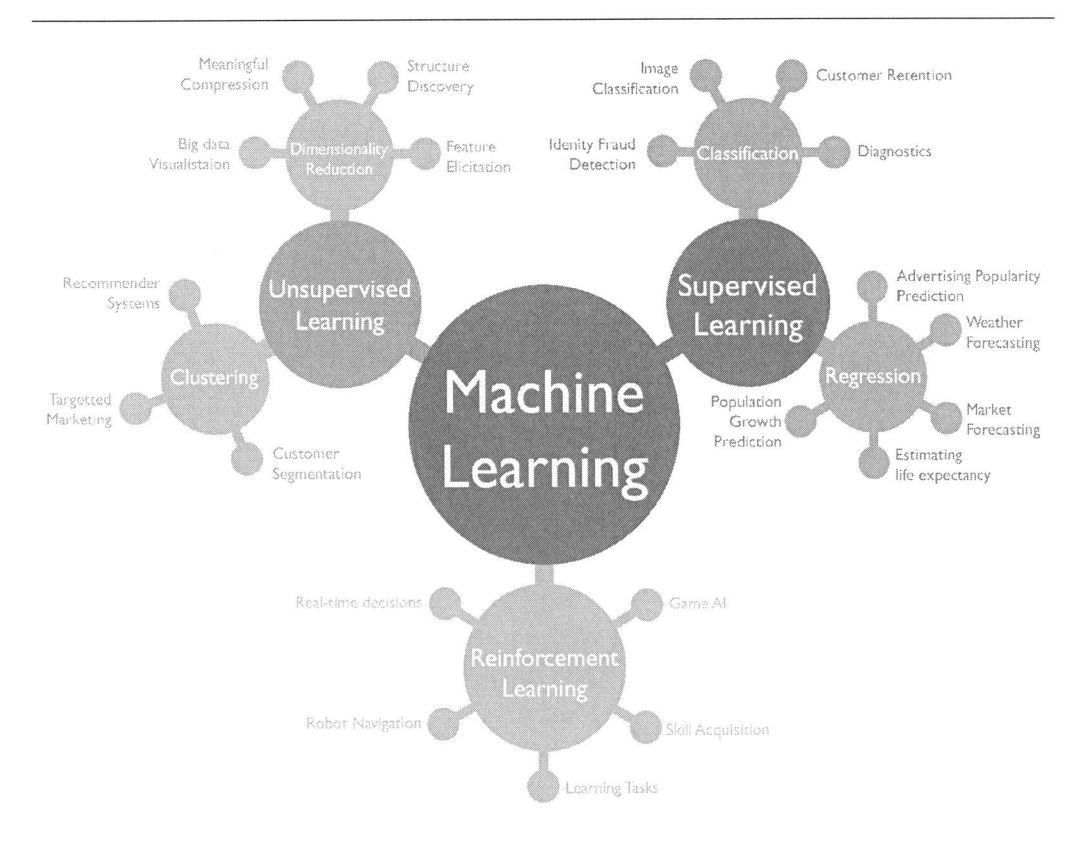

Figure 2.7: The three kinds of ML – supervised learning, unsupervised learning, and reinforcement learning

We will now talk about a few major domains where the ML model has safely secured its place as a predictive tool.

E-commerce

ML models help retailers understand the buying behavior of customers and their preferences. From historical purchase patterns of customers and click-through rates of products, e-commerce companies effectively recommend products and offer to maximize their sales. Personalized recommendations help retailers retain their customer base, thus creating loyalty. The following link outlines the particular ways ML can be utilized in the e-commerce industry:

`https://blog.shift4shop.com/machine-learning-ecommerce-industry`

Sales and marketing

ML models are used in B2B marketing as well. Identifying and acquiring prospects with features similar to existing businesses is one use case of customer segmentation. Prioritizing known prospects and generating new leads based on the likelihood of customers taking action is achieved using lead-scoring algorithms. Companies can streamline their sales and marketing activities by being data-driven as well as algorithm-driven. Here are some ways sales and marketing have improved when driven by ML:

```
https://scinapse.ai/blog/11-ways-machine-learning-can-improve-
marketing-and
```

Cybersecurity

Cyber-attacks may strike an organization at any time and cause serious harm; however, they can be predicted and prevented by ML algorithms. From processing both structured and unstructured data in a short time, real-time traffic can be analyzed to track unusual or anomalous patterns. Companies keep attacks at bay by analyzing these outlying points in the data. This also reduces the scope of human error stemming from the manual processing of massive volumes of data and enables humans to focus on strategizing the protection of the system from cyber-attacks. The following data-driven methods pointed out by Kaspersky are worth studying:

```
https://www.kaspersky.com/enterprise-security/wiki-section/products/
machine-learning-in-cybersecurity
```

Having explored how ML works as a predictive modeling tool in the industry, we will learn in the next section how mathematical modeling can be used as a prescriptive tool in different sectors.

Mathematical modeling – a prescriptive tool

Businesses often make complex decisions about their course of action to achieve objectives with the help of mathematical modeling or heuristics. A mathematical model in this sense is a prescriptive analytical tool. Answering the "where" and "when" is as important as answering what happened in the past (descriptive analytics) and what could happen in the future (predictive analytics). If a business wants to drive decisions from data in addition to insights and future predictions, it has to use both predictive and prescriptive tools in an integrated fashion.

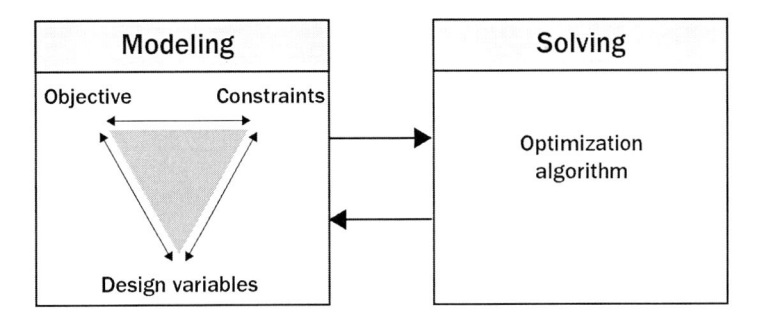

Figure 2.8: Mathematical optimization or mathematical modeling

We will have a look at examples from industry verticals wherein these work in tandem, resulting in higher productivity and profitability.

Finance

Financial services, including banks, rely on ML models as well as mathematical models to determine the right allocation of their investment portfolios. An ML model in the form of time-series forecasting helps with the prediction of asset performance, which in turn is channelled into applications leveraging a mathematical model. Based on the market movements and forecasts, the mathematical optimization application determines the optimal allocation. The best portfolio allocation also takes individual investment objectives and preferences into account. These mitigate risks and maximize risk-adjusted returns on investments.

Retail

Leading retailers utilize ML models to forecast demand for products, especially high-selling ones in particular locations at given times. They feed these predictions into mathematical models to maximize profits and customer satisfaction. The mathematical optimization application, in this case, uses the forecast as input to generate optimal production, pricing, inventory and distribution planning, logistics, and warehousing, thereby making the best business decisions while minimizing operating costs. Supply chain management is a classic example of mathematical optimization.

Energy

Governments and industry players are making high-stakes decisions on strategic investments in network infrastructure and resources as electric power is making a transition from being dependent on fossil fuels to renewables such as solar and wind. Organizations are utilizing ML models to predict future power demand and capacity needs. These forecasts are fed into mathematical models or mathematical optimization applications that generate optimal long-term investment planning and help in making decisions about strategic investments.

Digital advertising

Search engine giants such as Google leverage ML (and deep learning) models to predict the products and services individuals will be interested in looking up, based on their prior search history and a few other factors. In addition, they utilize mathematical models to figure out the online advertisements that can be shown to individual users at certain times. Search engine giants use this optimization model to charge advertisers and maximize their revenue.

These domains have added mathematical modeling to their data science toolbox that handles complex, significant, and scalable business problems for greater value delivery. Other industries, such as telecommunications and cloud computing, also use both models to precisely assess long-term demand and capacity needs to make the best business decisions.

Summary

In this chapter, we introduced ML models as problems of mathematical optimization or mathematical programming. We found out that an end-to-end ML project is the sum of multiple small optimization problems. We also gained knowledge about how businesses can unlock the true value of data upon leveraging mathematical models (primarily driven by mathematical equations) in addition to ML (driven by data) models. We learned that an ML model is predominantly a predictive tool and a mathematical model is a prescriptive one.

In the next chapter (which begins the next part of the book), we will take a meticulous look at a well-known algorithm called PCA, utilized in an unsupervised ML model fit to data with high dimensionality. It is a dimensionality reduction technique and one of the most tried and tested mathematical tools employing constrained optimization.

Part 2: Mathematical Tools

In this part, you will learn some of the most tried and tested mathematical tools and algorithms. On the one hand, there are algorithms for data dimensionality reduction, optimization of machine learning models, and data classification, which are explored through Python code. On the other hand, there are algorithms that model the relationships between objects (data points) and estimate the current and future states of variables (unknown and immeasurable ones) of a dynamic system. There are also other algorithms that predict the next future state probabilistically from knowledge of the present state of a process, explained with simple examples and Python code.

This part has the following chapters:

- *Chapter 3, Principal Component Analysis*
- *Chapter 4, Gradient Descent*
- *Chapter 5, Support Vector Machine*
- *Chapter 6, Graph Theory*
- *Chapter 7, Kalman Filter*
- *Chapter 8, Markov Chain*

3
Principal Component Analysis

A well-known algorithm to extract features from high-dimensional data for consumption in **machine learning** (**ML**) models is **Principal Component Analysis** (**PCA**). In mathematical terms, *dimension* is the minimum number of coordinates required to specify a vector in space. A lot of computational power is needed to find the distance between two vectors in high-dimensional space and in such cases, dimension is considered a curse. An increase in dimension will result in high performance of the algorithm only to a certain extent and will drop beyond that. This is the curse of dimensionality, as shown in *Figure 3.1*, which impedes the achievement of efficiency for most ML algorithms. The variable columns or features in data represent dimensions of space and the rows represent the coordinates in that space. With the increasing dimension of data, sparsity increases and there is an exponentially increasing computational effort required to calculate distance and density. Theoretically speaking, an increase in dimension practically increases noise and redundancy in large datasets. Arguably, PCA is the most popular technique to tackle this complexity of dimensionality in high-dimensional problems.

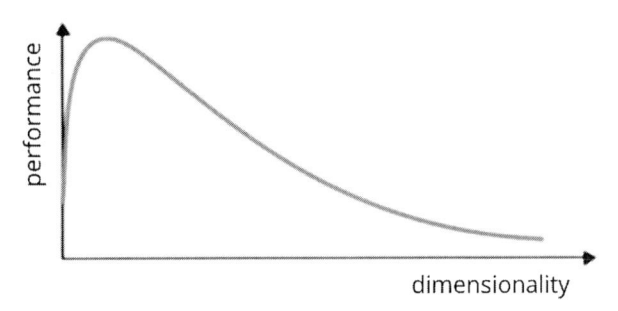

Figure 3.1: Curse of dimensionality

PCA comes from the field of linear algebra and is essentially a data preparation method that projects the data in a subspace before fitting the ML model to the newly created low-dimensional dataset. PCA is a data projection technique useful in visualizing high-dimensional data and improving data classification. It was invented in the 1900s following the principal axis theorem. The main objectives of PCA are to find an orthonormal basis for the data, sort dimensions in the order of importance or variance, discard dimensions of low importance, and focus only on uncorrelated Gaussian components.

This chapter covers the following topics related to PCA:

- Linear algebra for PCA
- Linear discriminant analysis – the difference from PCA
- Applications of PCA

The following section talks about linear algebra, the subject of mathematics on which PCA is based.

Linear algebra for PCA

PCA is an unsupervised method used to reduce the number of features of a high-dimensional dataset. An unlabeled dataset is reduced into its constituent parts by matrix factorization (or decomposition) followed by ranking of these parts in accordance with variances. The projected data representative of the original data becomes the input to train ML models.

PCA is defined as the orthogonal projection of data onto a lower dimensional linear space called the principal subspace, done by finding new axes (or basis vectors) that preserve the maximum variance of projected data; the new axes or vectors are known as principal components. PCA preserves the information by considering the variance of projection vectors: the highest variance lies on the first axis, the second highest on the second axis, and so forth. The working principle of the linear transformation called PCA is shown in *Figure 3.2*. It compresses the feature space by identifying a subspace that captures the essence of the complete feature matrix.

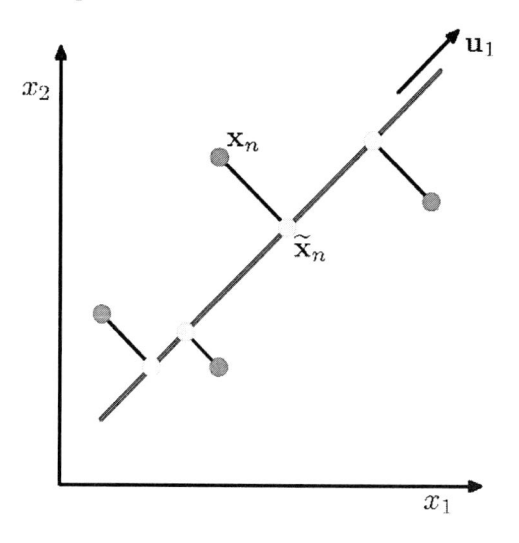

Figure 3.2: PCA working principle

There are other approaches to reducing data dimensionality such as feature selection methods (wrapper and filter), non-linear methods such as manifold learning (t-SNE), and deep learning (autoencoders) networks; however, the widest and most popular exploratory approach is PCA. Typically, linear algebraic methods assume that all inputs have the same distribution, and hence, it is a good practice to (either normalize or standardize) scale data before using PCA if the input features have different units.

Covariance matrix – eigenvalues and eigenvectors

The constraint in PCA is all the principal axes should be mutually orthogonal. The covariance of data is a measure of how much any pair of features in the data vary from each other. A covariance matrix checks the correlations between features in data and the directions of these relationships are obtained depending on whether the covariance is less than, equal to, or greater than zero. *Figure 3.3* displays the covariance matrix formula. Each element in the matrix represents the correlation between two features in the data where j and k run over p variates, N is the number of observations (rows), and the x_j bar and x_k bar in the formula denote the expected values (averages).

$$q_{jk} = \frac{1}{N-1} \sum_{i=1}^{N} (x_{ij} - \bar{x}_j)(x_{ik} - \bar{x}_k)$$

Figure 3.3: Covariance matrix

The objective of PCA is to explain most of the data variability with far fewer variables or features than that in the original dataset. Each of the N observations or records resides in p-dimensional space. Not all the dimensions are equally relevant. PCA seeks a smaller number of important dimensions, where importance is quantified by the amount of variation of the observations along each dimension. The dimensions figured out by PCA are a linear combination of the p features. We can take the linear combinations and reduce the number of plots required for visual analysis of the feature space while retaining the essence of the original data.

The eigenvalues and eigenvectors of a covariance matrix computed by eigendecomposition determine the magnitude and direction of the new subspace, respectively. In linear algebra, an **eigenvector** (associated with a set of linear equations) of a linear transformation is a non-zero vector that changes by a magnitude (scalar) when the transformation is applied to it. The corresponding **eigenvalue** is the magnitude or factor by which the eigenvector is scaled, and **eigendecomposition** is the factorization of a matrix into its eigenvectors and eigenvalues. The principal components are the eigenvectors. The top eigenvalues of a covariance matrix after sorting in descending order yield the principal components of the dataset.

The first **principal component** (**PC**) of data is the linear combination of the features that have the highest variance. The second PC is the linear combination of the features that have maximum variance out of all linear combinations uncorrelated with the first PC. The first two PCs are shown in *Figure 3.4*, and this computational process proceeds until all the PCs of the dataset are found. These PCs are essentially the eigenvectors, and linear algebraic techniques show that the eigenvector corresponding to the highest eigenvalue of the covariance matrix explains the greatest proportion of data variability. Each PC vector defines a direction in feature space and all of them are uncorrelated – that is, mutually orthogonal. The PCs form the basis of the new space.

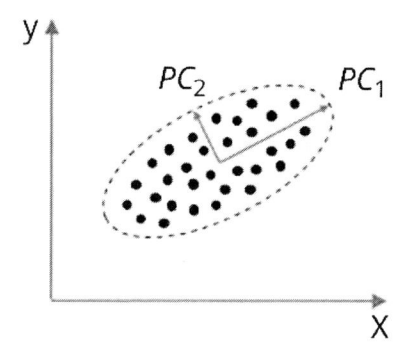

Figure 3.4: Principal components in feature space (x = x_1, y = x_2)

Number of PCs – how to select for a dataset

The question is how to determine how well a dataset is explained by a certain number of PCs. The answer lies in the percentage of variance retained by the number of PCs. We would ideally like to have the smallest number of PCs possible explaining most of the variability. There is no robust method to determine the number of usable components. As the number of observations and variables in the dataset vary, different levels of accuracy and amounts of reduction are desirable.

The **proportion of variance explained** (**PVE**) by the *m*th PC is computed by considering the *m*th eigenvalue. PVE is the ratio of the mth eigenvalue represented by ϕ_{jm} (of jth variate in *Figure 3.5*) and the sum of the eigenvalues of all the PCs or eigenvectors. To put it simply, PVE is the variance explained by each PC divided by the total variance of all PCs in the dataset.

$$PVE = \frac{\sum_{i=1}^{n}(\sum_{j=1}^{p} \phi_{jm} x_{ij})^2}{\sum_{j=1}^{p} \sum_{i=1}^{n} x_{ij}^2}$$

Figure 3.5: PVE calculation

In general, we look for the "elbow point" where the PVE significantly drops off to determine the number of usable PCs.

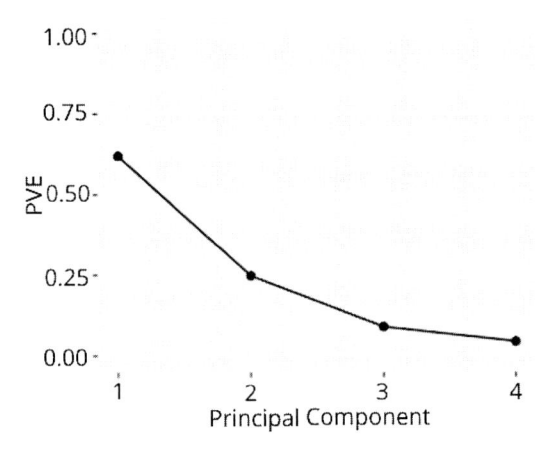

Figure 3.6a: PVE versus PC

The first PC in the example shown in *Figure 3.6a* explains 62.5% of data variability, and the second PC explains 25%. In a cumulative manner, the first two PCs explain 87.5% of the variability, as shown in *Figure 3.6b*.

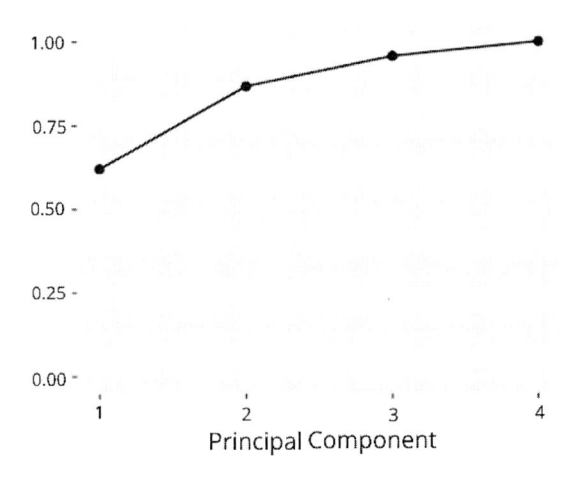

Figure 3.6b: Cumulative PVE (y axis) versus PC

Another widely used matrix decomposition method to identify the number of PCs is **singular value decomposition** (**SVD**). It is a reduced-rank approximation method that provides a simple means to do the same – that is, compute the principal components corresponding to the singular values. SVD is just another way to factorize a matrix and allows us to unravel similar information as eigendecomposition does. The singular values of the data matrix obtained via SVD are essentially the square roots of the eigenvalues of the covariance matrix in PCA. SVD is the same as PCA of a raw data matrix, which is mean-centered.

A mathematical demonstration of SVD can be found on the Wolfram pages: https://mathworld.wolfram.com/SingularValueDecomposition.html.

SVD is an iterative numerical method. Every rectangular matrix has SVD, although, for a few complex problems, it may fail to decompose some matrices neatly. You can perform SVD using the linear algebra class of the Python library, scipy. The scikit-learn library provides functions for SVD: https://scikit-learn.org/stable/modules/generated/sklearn.decomposition.PCA.html.

High-dimensional data can be reduced to a subset of dimensions (columns) that are most relevant to the problem being solved. The data matrix (rows by columns) results in a matrix with a lower rank that approximates the original matrix and best captures its salient features.

Feature extraction methods

There are two ways in which dimensionality reduction can be done: one is feature selection and the other is feature extraction. A subset of original features is selected in the former approach by filtering based on some criteria true to the particular use case and corresponding data. On the other hand, a set of new features is found in the feature extraction approach.

Feature extraction is done using linear mapping from the original features, which no longer exist upon implementation of the method. In essence, new features constructed from available data do not have column names as in the original data. There are two feature extraction methods: PCA and LDA. A nonlinear mapping may also be used depending on the data but the method no longer remains PCA or LDA.

Now that we have explored PCA for the reduction of features (and hence, the reduction of the high-dimensional space), we shall learn about a supervised method of linear feature extraction called **linear discriminant analysis** (**LDA**).

LDA – the difference from PCA

LDA and PCA are linear transformation methods; the latter yields directions or PCs that maximize data variance and the former yields directions that maximize the separation between data classes. The way in which the PCA algorithm works disregards class labels.

LDA is a supervised method to reduce dimensionality that projects the data onto a subspace in a way that maximizes the separability between (groups) classes; hence, it is used for pattern classification problems. LDA works well for data with multiple classes; however, it makes assumptions of normally distributed classes and equal class covariances. PCA tends to work well if the number of samples in each class is relatively small. In both cases, though, observations ought to be much higher relative to the dimensions for meaningful results.

LDA seeks a projection that discriminates data in the best possible way, unlike PCA, which seeks a projection that preserves maximum information in the data. When regularization of the estimate of covariance is introduced to moderate the influence of different variables on LDA, it is called **regularized discriminant analysis**.

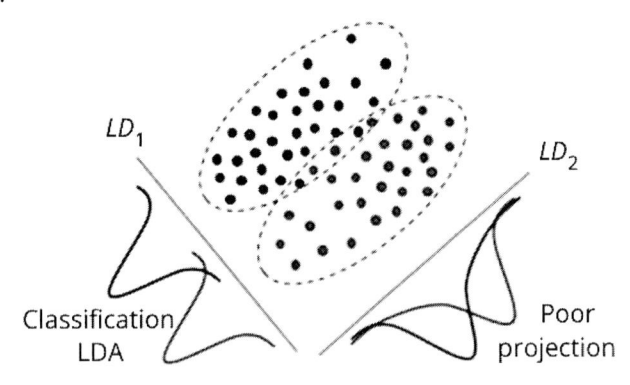

Figure 3.7: Linear discriminants

LDA involves developing a probabilistic model per class based on the distribution of each input variable (*Figure 3.7*). It may be considered as an application of the Bayes' theorem for classification and assumes that the input variables are uncorrelated; if they are correlated, the PCA transform may aid in removing the linear dependence. The scikit-learn library provides functions for LDA. Example code with a synthetic dataset is given here:

```
from sklearn.datasets import make_classification
from sklearn.model_selection import GridSearchCV
from sklearn.model_selection import RepeatedStratifiedKFold
from sklearn.discriminant_analysis import LinearDiscriminantAnalysis

X, y = make_classification(n_samples = 1000, n_features = 8, n_
informative = 8,
n_redundant = 0, random_state = 1) #train examples and labels

model = LinearDiscriminantAnalysis()
cv = RepeatedStratifiedKFold(n_splits = 10, n_repeats = 3, random_
state = 1)

grid = dict()
grid['solver'] = ['svd', 'lsqr', 'eigen'] #grid configuration
search = GridSearchCV(model, grid, scoring = 'accuracy', cv = cv, n_
jobs = -1)
results = search.fit(X, y)
print('Mean Accuracy: %.4f' % results.best_score_) #model accuracy
check
```

```
row = [0.1277, -3.6440, -2.2326, 1.8211, 1.7546, 0.1243, 1.0339,
2.3582] #new example
yhat = search.predict([row]) #predict on test data
print('Predicted Class: %d' % yhat) #class probability of new example
```

In the preceding example, the hyperparameter (`solver`) in the grid search is set to `'svd'` (default) but other `solver` values can also be used. This example only introduces us to using LDA with scikit-learn; there is a whole lot of customization that can be done depending on the problem being solved.

We have explored the linear algebraic methods for dimensionality reduction; we shall learn about the most important applications of PCA in the next section.

Applications of PCA

PCA is one fundamental algorithm and forms the foundation of ML. It finds use in diverse areas such as noise reduction in images, classification of data in general, anomaly detection, and other applications in medical data correlation. We will explore a couple of widely used applications of PCA in this section.

The scikit-learn library in Python provides functions for PCA. The following example code shows how to leverage PCA for dimensionality reduction while developing a predictive model that uses a PCA projection as input. We will be using PCA on a synthetic dataset while fitting a logistic regression model for classification:

```
from sklearn.datasets import make_classification
from sklearn.model_selection import cross_val_score
from sklearn.model_selection import RepeatedStratifiedKFold
from sklearn.pipeline import Pipeline
from sklearn.decomposition import PCA
from sklearn.linear_model import LogisticRegression
from numpy import mean
from numpy import std
import matplotlib.pyplot as plt

def get_models():
    models = dict()
    for i in range(1, 11):
        steps = [('pca', PCA(n_components = i)), ('m',
LogisticRegression())]
models[i] = Pipeline(steps = steps)
return models
def evaluate_model(model, X, y):
    cv = RepeatedStratifiedKFold(n_splits = 10, n_repeats = 3,
random_state = 1)
scores = cross_val_score(model, X,  y, scoring = 'accuracy', cv = cv,
```

```
            n_jobs = -1, error_score = 'raise')
    return scores
X, y = make_classification(n_samples = 1000, n_features = 10, n_
informative = 8, n_redundant = 2, random_state = 7)
models = get_models()
results, names = list(), list()
for name, model in models.items():
    scores = evaluate_model(model, X, y)
    results.append(scores)
    names.append(name)
print('Mean Accuracy: %.4f (%.4f)' % (mean(results), std(results)))
red_square = dict(markerfacecolor = 'r', marker = 's')
```

We'll plot the grid now:

```
plt.boxplot(results, labels = names, showmeans = True, showfliers =
True, flierprops = red_square)
plt.grid()
plt.xlabel('Principal Components')
plt.xticks(rotation = 45)
plt.show()
row = [0.1277, -3.6440, -2.2326, 1.8211, 1.7546, 0.1243, 1.0339,
2.3582, -2.8264,0.4491] #new example
steps = [('pca', PCA(n_components = 8)), ('m', LogisticRegression())]
model = Pipeline(steps = steps)
model.fit(X, y)
yhat = model.predict([row]) #predict on test data
print('Predicted Class: %d' % yhat) #predicted class of new example
```

In the preceding example, we do not see improvement in the model accuracy beyond eight components (*Figure 3.8*). It is evident that the first eight components contain maximum information about the class and the remaining ones are redundant.

38   Principal Component Analysis

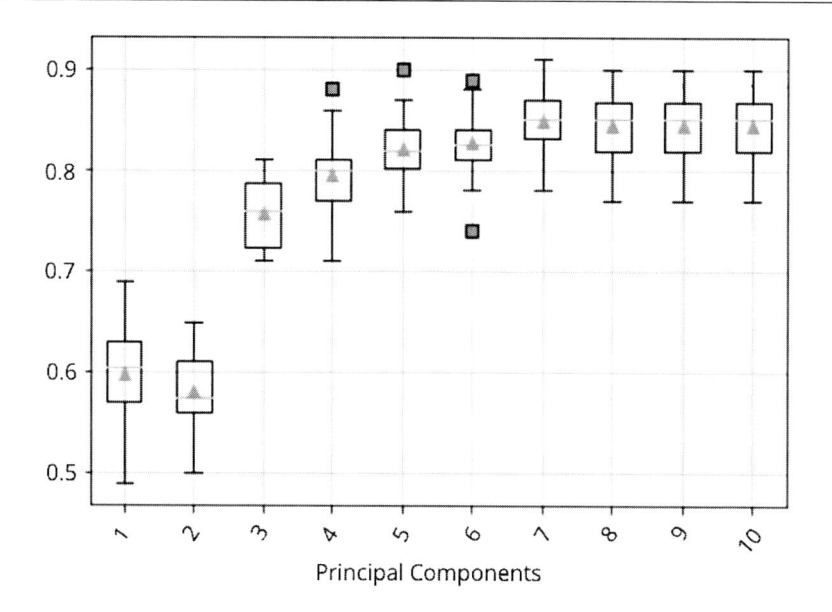

Figure 3.8: Classification accuracy versus the number of PCs for a synthetic dataset

The number of components after a PCA transform of features that results in the best average performance of the model is chosen and fed to the ML model for predictions. In the following subsection, we will learn about denoising and the detection of outliers using PCA.

Noise reduction

PCA finds use in the reduction of noise in data, especially images. PCA reconstruction of an image by denoising can be achieved with the decomposition method of the scikit-learn Python library. Details of the library function with examples can be found here: https://scikit-earn.org/stable/auto_examples/applications/plot_digits_denoising.html

A good exercise would be to reconstruct images obtained from a video sequence exploring linear PCA as well as kernel PCA and check which one provides smoother images.

Image compression (*Figure 3.9*) is another important application of PCA.

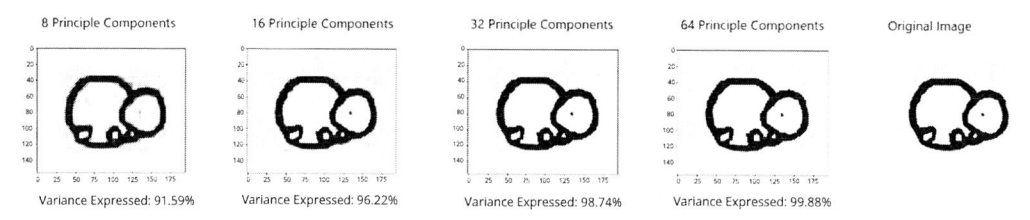

Figure 3.9: Image compression with PCA

The percentage of variance expressed by the PCs determines how many features should be the input to deep learning models (neural networks) for image classification so that the computing performance is not affected while dealing with huge and high-dimensional datasets.

Anomaly detection

Detecting anomalies is common in fraud detection, fault detection, and system health monitoring in sensor networks. PCA makes use of the cluster method for detecting an outlier, typically collective and unordered outliers. Cluster-based anomaly detection assumes that the inlying (normal) data points belong to large and dense clusters, and outlying (anomalous) ones belong to small or sparse clusters or do not belong to any of them. Example code with sample telemetry data can be found in the following repository: `https://github.com/ranja-sarkar/mm`.

The PCs apply distance metrics to identify anomalies. PCA, in this case, determines what constitutes a normal class. As an exercise, you can use unsupervised learning methods such as K-means and Isolation Forest to detect outliers in the same dataset for a comparison of the results and gain more meaningful insights.

Summary

In this chapter, we learned about two linear algebraic methods used to reduce the dimensionality of data: namely, principal component analysis and linear discriminant analysis. The focus was on PCA, which is an unsupervised method to reduce the feature space of high-dimensional data and to know why this reduction is necessary for solving business problems. We did a detailed study of the mathematics behind the algorithm and how it works in ML models. We also learned about a couple of important applications of PCA along with the Python code.

In the next chapter, we will learn about an optimization method called Gradient Descent, which is arguably the most common (and popular) algorithm to optimize neural networks. It is a learning algorithm that works by minimizing a given cost function. As the name suggests, it uses a gradient (derivative) iteratively to minimize the function.

4
Gradient Descent

One optimization algorithm that lays the foundation for machine learning models is **gradient descent** (**GD**). GD is a simple and effective tool useful to train such models. Gradient descent, as the name suggests, involves "going downhill." We choose a direction across a landscape and take whichever step gets us downhill. The step size depends on the slope (gradient) of the hill. In **machine learning** (**ML**) models, gradient descent estimates the error gradient, helping to minimize the cost function. Very few optimization methods are as computationally efficient as gradient descent. GD also lays the foundation for the optimization of deep learning models.

In problems where the parameters cannot be calculated analytically by use of linear algebra and must be searched by optimization, GD finds its best use. The algorithm works iteratively by moving in the direction of the steepest descent. At each iteration, the model parameters, such as coefficients in linear regression and weights in neural networks, are updated. The model continues to update its parameters until the cost function converges or reaches its minimum value (the bottom of the slope in *Figure 4.1a*).

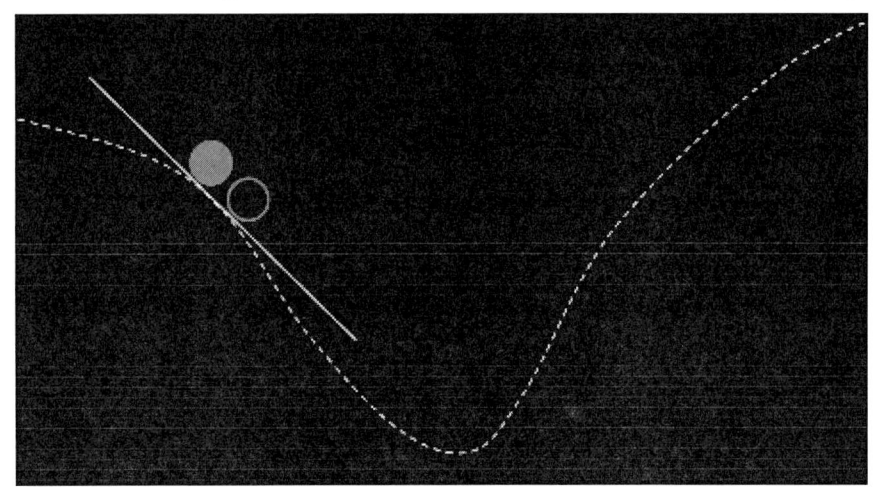

Figure 4.1a: Gradient descent

The size of a step taken in each iteration is called the learning rate (a function derivative is scaled by the learning rate at each iteration). With a learning rate that is too low, the model may reach the maximum permissible number of iterations before reaching the bottom, whereas it may not converge or may diverge (the so-called exploding gradient problem) completely if the learning rate is too high. Selecting the most appropriate learning rate is crucial in achieving a model with the best possible accuracy, as seen in *Figure 4.1b*.

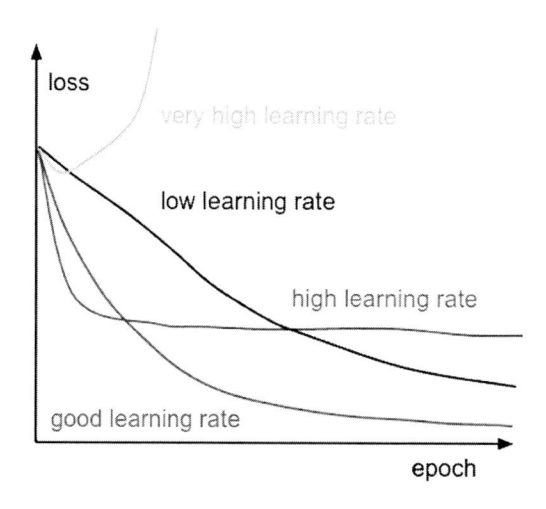

Figure 4.1b: Learning rates in gradient descent

For GD to work, the objective or cost function must be differentiable (meaning the first derivative exists at each point in the domain of a univariate function) and convex (where two points on the function can be connected by a line segment without crossing). The second derivative of a convex function is always positive. Examples of convex and non-convex functions are shown in *Figure 4.2*. GD is a first-order optimization algorithm.

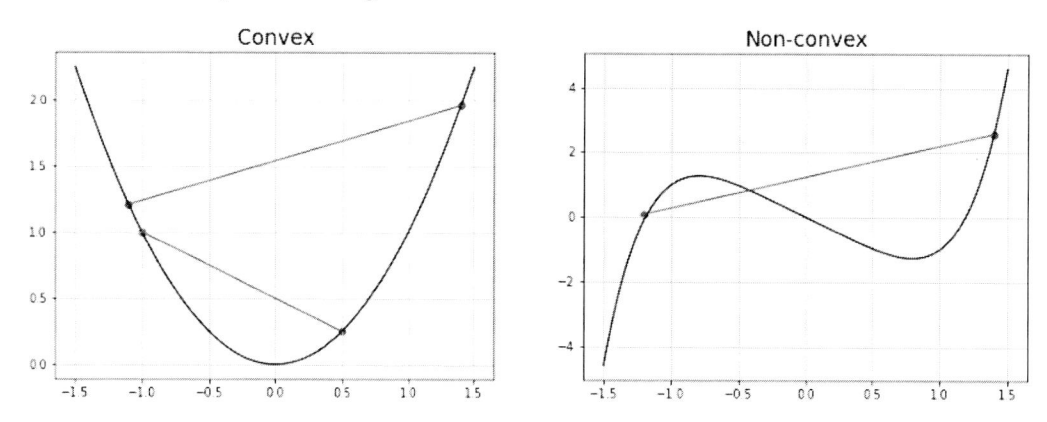

Figure 4.2: Example of convex (L) and non-convex (R) function

In a multivariate function, the gradient is a vector of derivatives in each direction in the domain. Such functions have saddle points (quasi-convex or semi-convex) where the algorithm may get stuck and obtaining a minimum is not guaranteed. This is where second-order optimization algorithms are brought in to escape the saddle point and reach the global minimum. The GD algorithm finds its use in control as well as mechanical engineering, apart from ML and DL. The following sections compare the algorithm with other optimization algorithms used in ML and **deep learning** (**DL**) models and specifically examines some commonly used gradient descent optimizers.

This chapter covers the following topics:

- Gradient descent variants
- Gradient descent optimizers

Gradient descent variants

The workings of the gradient descent algorithm to optimize a simple linear regression model ($y = mx + c$) is elaborated with Python code in this section.

Application of gradient descent

Keeping the number of iterations the same, the algorithm is run for three different learning rates resulting in three models, hence three **MSE** (**mean squared error**) values. MSE is the calculated loss or cost function in linear regression:

```
import numpy as np
import matplotlib.pyplot as plt
from sklearn.metrics import mean_squared_error
#gradient descent method
class GDLinearRegression:
    def __init__(self, learning_rate, epoch):
        self.learning_rate, self.iterations = learning_rate, epoch
        #epoch is number of iterations
    def fit(self, X, y):
        c = 0
        m = 5
        n = X.shape[0]
        for _ in range(self.iterations):
            b_gradient = -2 * np.sum(y - m*X + c) / n
            m_gradient = -2 * np.sum(X*(y - (m*X + c))) / n
            c = c + (self.learning_rate * b_gradient)
            m = m - (self.learning_rate * m_gradient)
        self.m, self.c = m, c
    def predict(self, X):
```

```
            return self.m*X + self.c
#dataset
np.random.seed(42)
X = np.array(sorted(list(range(5))*20)) + np.random.normal(size = 100,
scale = 0.5)
y = np.array(sorted(list(range(5))*20)) + np.random.normal(size = 100,
scale = 0.3)
#model 1
Clf_1 = GDLinearRegression(learning_rate = 0.05, epoch = 1000)
Clf_1.fit(X, y)
y_pred = Clf_1.predict(X)
mse_1 = mean_squared_error(y, y_pred)
plt.style.use('fivethirtyeight')
plt.scatter(X, y, color='black')
plt.plot(X, y_pred)
plt.gca().set_title("Linear Regression Model 1")
print('Slope = ', round(Clf_1.m, 4))
print('Intercept = ', round(Clf_1.c, 4))
print('MSE = ', round(mse_1, 2))
```

Two other models are trained with two different learning rates, one higher and another lower than model 1, as seen here:

```
#model 2
Clf_2 = GDLinearRegression(learning_rate = 0.2, epoch = 1000)
Clf_2.fit(X, y)
y_pred = Clf_2.predict(X)
mse_2 = mean_squared_error(y, y_pred)
plt.style.use('fivethirtyeight')
plt.scatter(X, y, color='black')
plt.plot(X, y_pred)
plt.gca().set_title("Linear Regression Model 2")
print('MSE = ', round(mse_2, 2))
#model 3
Clf_3 = GDLinearRegression(learning_rate = 0.0001, epoch = 1000)
Clf_3.fit(X, y)
y_pred = Clf_3.predict(X)
mse_3 = mean_squared_error(y, y_pred)
plt.style.use('fivethirtyeight')
plt.scatter(X, y, color='black')
plt.plot(X, y_pred)
plt.gca().set_title("Linear Regression Model 3")
print('MSE = ', round(mse_3, 2))
```

Upon executing the code, the linear regression models obtained (*Figure 4.3*) show how carefully the parameter (learning rate) should be chosen to attain optimal performance or the best accuracy of the ML model.

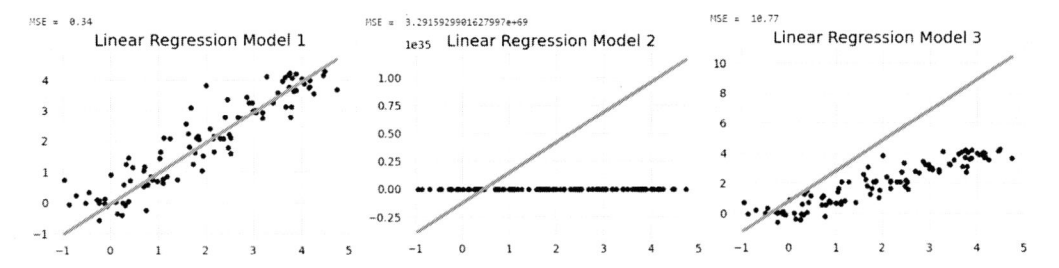

Figure 4.3: Gradient descent for a linear regression model

There are GD variants (*Figure 4.4*) that differ in the data size used to compute the gradient of the objective function. A trade-off between the accuracy of the parameter (coefficient or weight) and the time taken to do it is made depending on the amount of data. The variants are **batch gradient descent** (**BGD**), **mini-batch gradient descent**, and **stochastic gradient descent** (**SGD**), which we will now discuss in the following subsection.

Mini-batch gradient descent and stochastic gradient descent

BGD, also known as vanilla gradient descent, is simply gradient descent and computes the gradient for the entire training data to perform one update (in one step) and thus can be very slow. Common examples of ML models that are optimized using BGD are linear regression and logistic regression for smaller datasets.

For bigger datasets, we generally use mini-batch GD, which allows the splitting of training data into mini-batches that can be processed individually. After each mini-batch is processed, the parameters are updated and this continues until the entire dataset has iteratively been processed. One full cycle through the data is called an epoch. A number of steps are taken to reach the global minimum, which introduces some variance into the optimization process. This variant of GD is usually used for modeling problems where efficiency is as important as accuracy.

SGD performs frequent parameter updates (for each training example) with a high level of variance that causes the cost function to fluctuate heavily. This enables it to jump to a new and potentially better local minimum. Upon slowly decreasing the learning rate, SGD shows convergence behavior similar to BGD. SGD is computationally faster than BGD as it considers one example at a time.

46   Gradient Descent

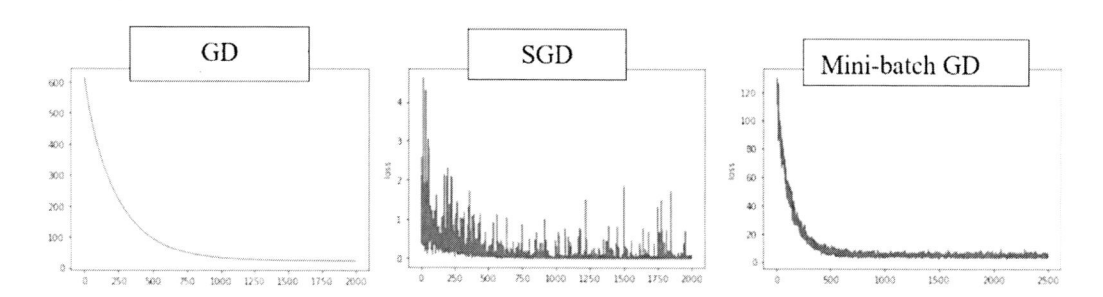

Figure 4.4: Gradient descent variants

SGD is typically the algorithm of choice for training a neural network. A key challenge with SGD while minimizing the highly non-convex error functions common in neural networks is avoiding getting trapped in their numerous suboptimal local minima. These saddle points make it very hard for SGD to escape, as the gradient is close to zero in all dimensions. In the next section, we outline some GD optimizers that deal with such challenges.

Gradient descent optimizers

The optimizers discussed here are widely used to train DL models depending on the degree of the non-convexity of the error or cost function.

Momentum

The momentum method uses a moving average gradient instead of a gradient at each time step and reduces the back-and-forth oscillations (fluctuations of the cost function) caused by SGD. This process focuses on the steepest descent path. *Figure 4.5a* shows movement with no momentum by creating oscillations in SGD while *Figure 4.5b* shows movement in the relevant direction by accumulating velocity with damped oscillations and closer to the optimum.

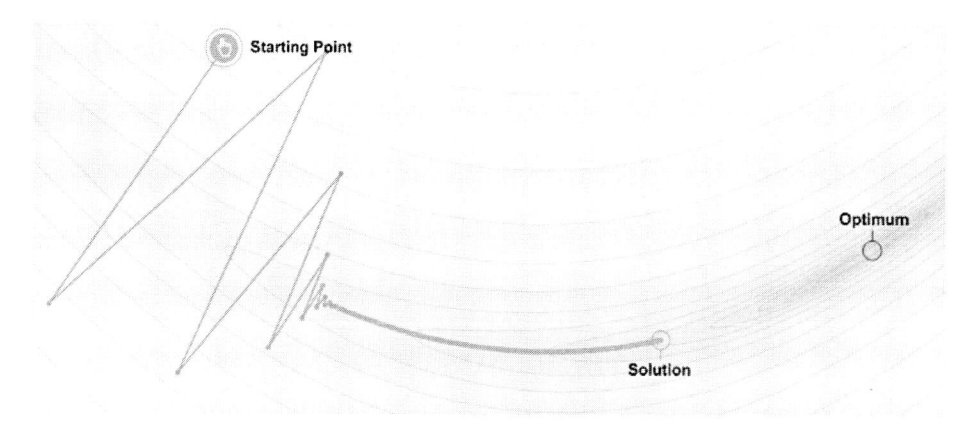

Figure 4.5a: SGD with no momentum

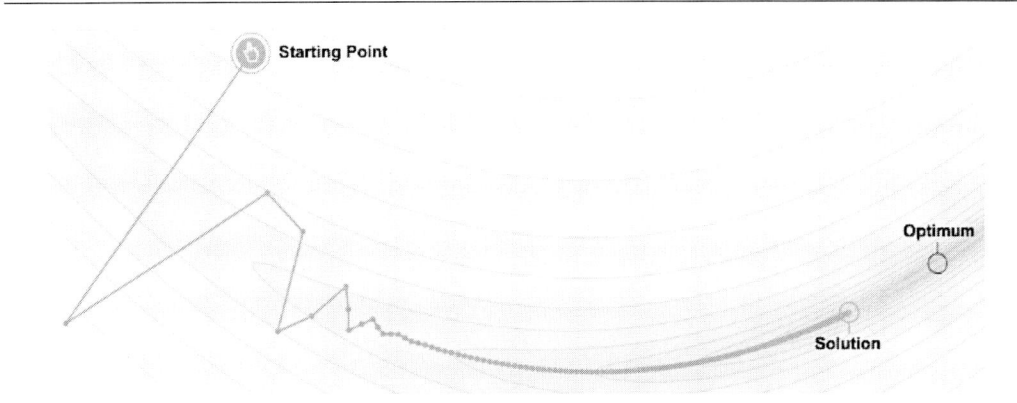

Figure 4.5b: SGD with momentum

The momentum term reduces updates for dimensions whose gradients change directions and as a result, faster convergence is achieved.

Adagrad

The `adagrad` optimizer is used when dealing with sparse data as the algorithm performs small updates of parameters based on features that occur often. In `adagrad`, different or "adaptive" learning rates are used for every update at every time step (*Figure 4.6*). The algorithm uses larger learning rates for infrequent features and smaller ones for more frequent features. The major advantage of using this optimizer is that the learning rate is not set manually. And when the learning rate shrinks to almost zero, the model gains no new knowledge.

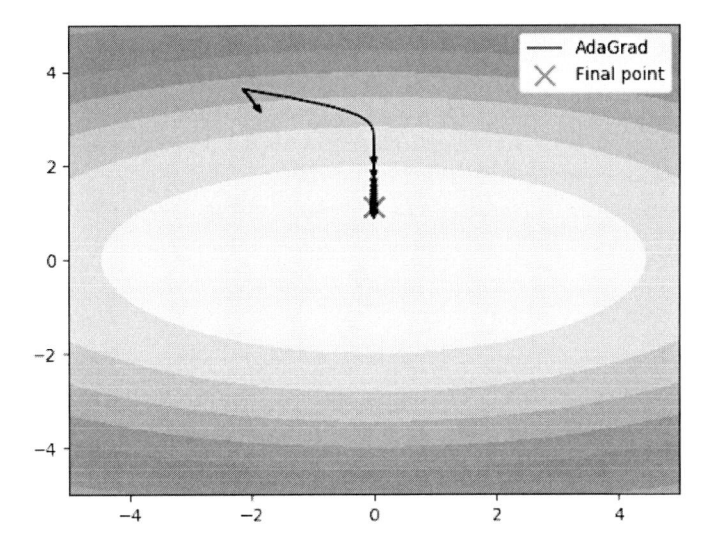

Figure 4.6: Adagrad optimizer

RMSprop

The RMSprop optimizer is similar to the `adagrad` optimizer and hence known as **leaky adagrad**, only it uses a different method for parameter updates. The RMSprop algorithm restricts oscillations in the vertical direction so that it can take larger steps in the horizontal direction (*Figure 4.7*). The algorithm adaptively scales the learning rate in each dimension by using an exponentially weighted average of the gradient that allows it to focus on the most recent gradients.

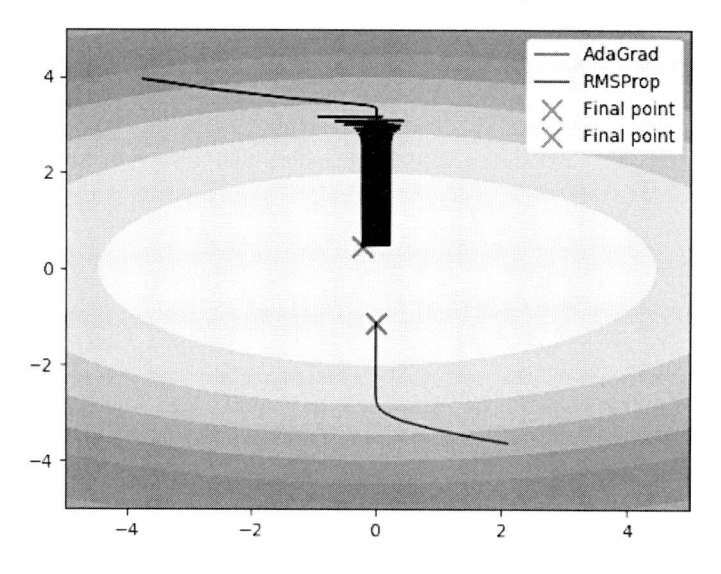

Figure 4.7: RMSprop optimizer

Adam

The **adaptive moment estimation (adam)** optimizer inherits the advantages of both the momentum and RMSprop optimization algorithms (*Figure 4.8*). It combines the ideas of a moving average gradient and an adaptive learning rate. These two respectively represent the estimates of the first moment (mean) and second moment (variance) of the gradient of the cost function, hence the name.

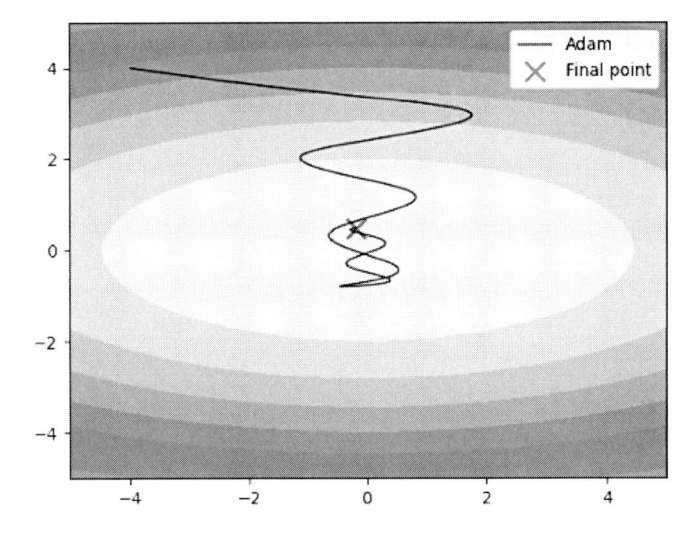

Figure 4.8: Adam optimizer

It has been empirically observed that the adam optimizer is effective and works better than SGD in practice. It has become the default optimizer of choice to train DL models. For further reading, check out the following MachineLearningMastery article: `https://machinelearningmastery.com/adam-optimization-algorithm-for-deep-learning/`.

Summary

In this chapter, we learned about a foundational optimization algorithm and its variants used in training ML and DL models. An application of the optimization technique in Python to a linear regression problem was also elaborated on. Both the cost function and its gradient, and how to update the gradient to converge to the optimal point, are mathematical concepts every data scientist must understand thoroughly; optimizing a cost function is the basis of achieving an optimal model for a problem or predictions. Different ways can be used to estimate the gradients depending on the behavior of the cost function.

In the following chapter, we will explore another fundamental algorithm, known as **support vector machines** (**SVMs**). Although SVMs can be used for regression problems, they are more widely used for classification tasks.

5
Support Vector Machine

This chapter explores a classic algorithm that one must keep in one's machine learning arsenal called the **support vector machine** (**SVM**), which is mainly used for classification problems rather than regression problems. Since its inception in the 1990s, it was commonly used to recognize patterns and outliers in data. Its popularity declined after the emergence of boosting algorithms such as **extreme gradient boost** (**XGB**). However, it prevails as one of the most commonly used supervised learning algorithms.

In the 1990s, efficient learning algorithms based on computational learning were developed for non-linear functions. Algorithms such as linear learning algorithms have well-defined theoretical properties. With this development, efficient separability (decision surfaces) of nonlinear regions that use kernel functions was established. Nonlinear SVMs are quite frequently used for the classification of real (nonlinear) data.

SVM was initially known as a binary classifier that could be used for one-class classification of skewed or imbalanced class distribution. This unsupervised algorithm could effectively learn from the majority or normal class in a dataset to classify new data points as either *normal* or *outlier*. The process of identifying the minority or rarity class generally referred to as outlier is called **anomaly detection**, as the outlier is an anomaly and the rest of the data is normal. Classification involves fitting a model on the normal data (training examples) and predicting whether incoming new data is normal (inlier) or outlier. One-class SVM is most suited for a specific problem where the minority class does not have a consistent pattern or is a noisy instance, making it difficult for other classification algorithms to learn a decision boundary. The outliers in general are treated as deviations from normal.

In general, SVMs are effective in problems where the number of variables is greater than the number of records, meaning, in high-dimensional spaces. The algorithm uses a subset of training examples in the decision function, hence it is memory-efficient. It turns out the algorithm is versatile, as different kernels can be specified for the decision function.

This chapter covers the following topics:

- Support vectors in SVM
- Kernels for SVM
- Implementation of SVM

We will learn about support vectors and kernels in the forthcoming sections.

Support vectors in SVM

SVM is an algorithm that can produce significantly accurate results with less computation power. It is widely used in data classification tasks. If a dataset has n number of features, SVM finds a hyperplane in the n-dimensional space, which is also called the **decision boundary**, to classify the data points. An optimal decision boundary maximizes the distance between the boundary and instances in both classes. The distance between data points in the classes (shown in *Figure 5.1a*) is known as the **margin**:

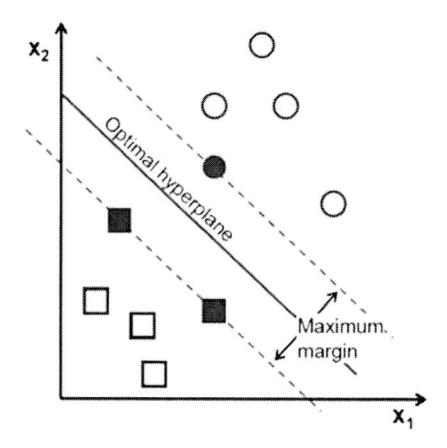

Figure 5.1a: Optimal hyperplane

An SVM algorithm finds the optimal line in two dimensions or the optimal hyperplane in more than two dimensions that separates the space into classes. The optimal hyperplane or optimal line maximizes the margin (the distance between the data points of the two classes). In 3D (or more), data points become vectors and those (very small subset of training examples) that are closest to or on the hyperplanes (just outside the maximum margin) are called **support vectors** (see *Figure 5.1b*):

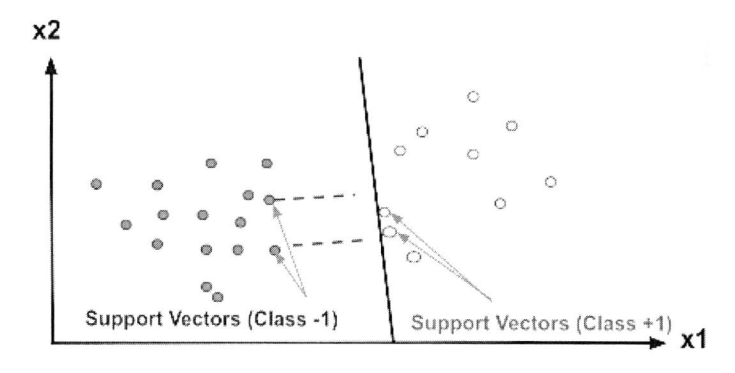

Figure 5.1b: Support vectors

If all support vectors are at the same distance from the optimal hyperplane, the margin is said to be good. The margin shown in *Figure 5.1b* is bad, as support vectors in class +1 are very close to the optimal hyperplane, while those in class -1 are far away from it. Moving a support vector moves the decision boundary or hyperplane while moving other data points has no effect.

If the number of features in an input dataset is two, the hyperplane is just a line. If the number is three, then the hyperplane (shown in *Figure 5.2*) is a two-dimensional plane. The dimension of the decision boundary depends on the number of features, and the data points on either side of it (the hyperplane) belong to different classes:

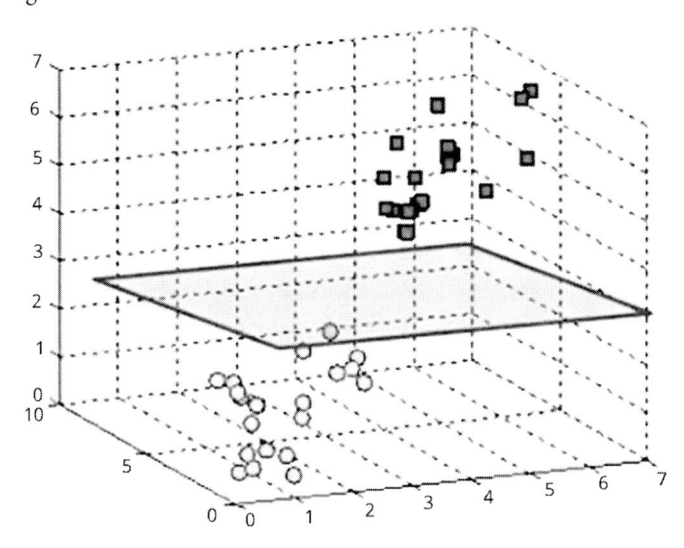

Figure 5.2: Hyperplane in 3D feature space

Support vectors influence the position and orientation of the hyperplane. The margin of the classifier is maximized using support vectors. The margin is hard if the data is linearly separable. For most practical problems, data is not linearly separable, and in such cases the margin is soft. This allows for data points within the marginal distance (shown in *Figure 5.3*) between two data class separators:

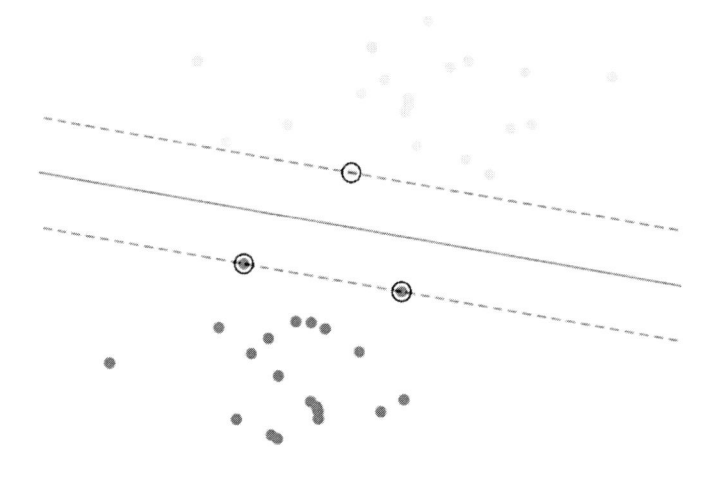

Figure 5.3: Soft margin

It is better to have a large margin that might allow for some margin violation to occur. The larger the margin, the lower the error of the classifier. Maximizing the margin is equivalent to minimizing loss in machine learning algorithms. The function that helps maximize the margin is **hinge loss**. Hinge loss (error) is zero if data is classified correctly, meaning we have a hard margin as the points are not close to the hyperplane. Hinge loss is one if most of the data points are classified incorrectly. In general, support vectors are within the margin boundaries (soft margin) when the problem is not linearly separable.

In the next section, the kernel trick is introduced. Kernel is a technique used in SVMs to classify data points that are not linearly separable. Kernel functions enable operation in a high-dimensional feature space without computing data coordinates in that space, hence this operation is not computationally expensive.

Kernels for SVM

With a kernel trick, a 2D space is converted into a 3D space using a mapping function such that the nonlinear data can be classified or separated in a higher dimension (see *Figure 5.4*). The transformation of original data for mapping into the new space is done via kernel. The kernel function defines inner products (measure of similarity) in the transformed space.

The compute and storage requirements of SVMs increase with the number of training examples. The core of the algorithm is a quadratic programming problem separating support vectors from the training dataset. A linear kernel, which is just a dot product, is the fastest implementation of SVM. A few examples of linear and nonlinear kernels are shown in *Figure 5.5a*. The most common nonlinear SVM kernels are **radial basis function** (**RBF**), **sigmoid**, and **polynomial**.

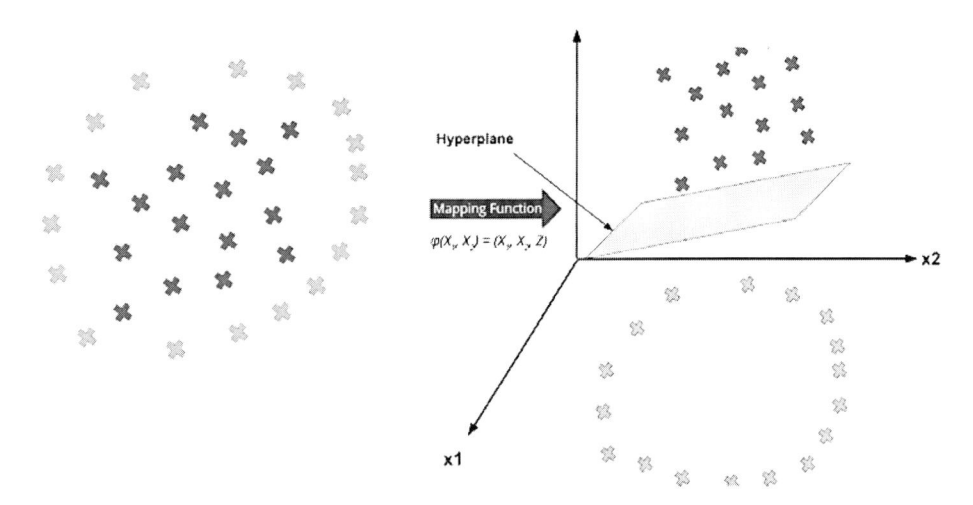

Figure 5.4: (a) Example of non-linear separator (L), and (b) Data effectively classified in higher dimension

SVMs are very effective for small datasets that are not linearly separable. Small data means that the number of features is more than the training size, due to which SVMs suffer from overfitting in some cases. The right kernel function and regularization (penalty function) come to the rescue in those cases. Each kernel has a different mathematical formulation, hence the set of parameters varies from one to another.

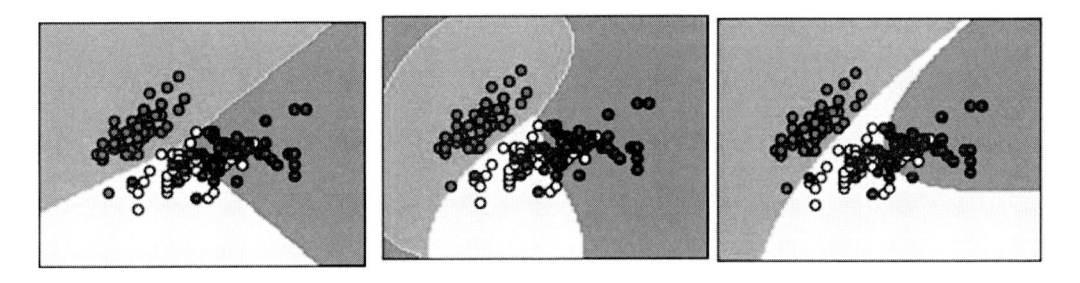

Figure 5.5a: Data classification using linear kernel (L), RBF kernel (M), and polynomial kernel (R) functions

The parameter exponent (degree) in a polynomial kernel when set to 1 becomes a linear kernel and when set to 3 becomes a cubic kernel, an example of which is shown in *Figure 5.5a* (rightmost). The sigmoid kernel cumulative distribution function goes from 0 to 1 to classify data and is mostly used as an activation function or perceptron in neural networks. An example of data classification by SVM with sigmoid kernel function is shown in *Figure 5.5b*:

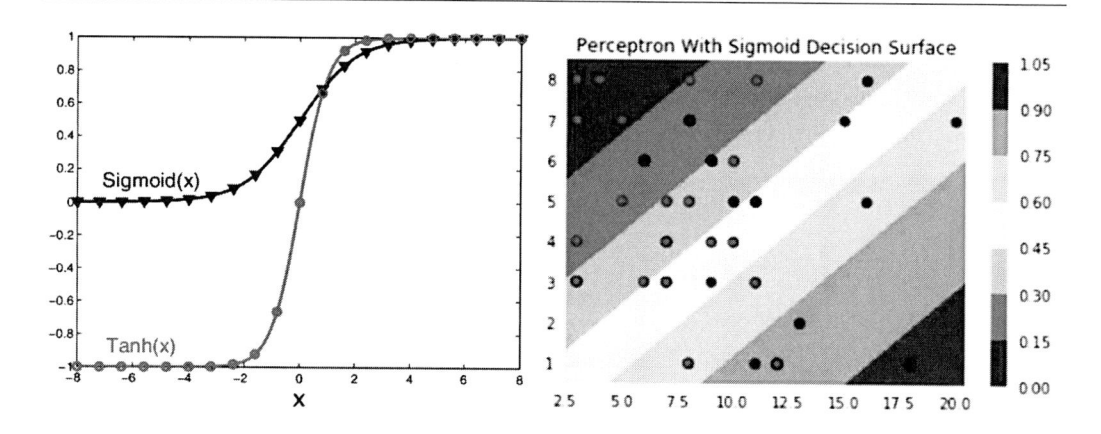

Figure 5.5b: Sigmoid kernel (L), data classified using sigmoid (R)

All SVM kernels have a parameter that trades off the misclassification of the dataset against the simplicity of the separator. While training an SVM with the RBF kernel, which is an exponential (e^{-ax^2}) function, the parameter a is greater than zero and defines the influence of a training example on the separator. The selection of the parameters in respective kernel functions is critical to an SVM's performance.

In imbalanced datasets, the parameters dedicated to providing weights on classes and samples become significant, as they might be required to give more importance to a certain sample or class in such cases. The effect of sample weighting on the class boundary is shown in *Figure 5.6*, wherein the data point size is proportional to the sample weights:

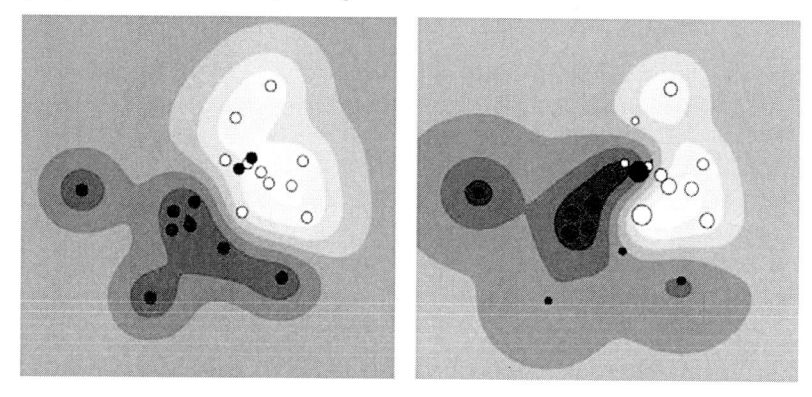

Figure 5.6: Classification with constant sample weight (L), with modified weight (R)

While various methods and algorithms can detect outliers in a dataset, the kernel method used by the one-class SVM algorithm has been demonstrated in this chapter. Other examples include the decision tree ensemble method in the Isolation Forest algorithm, the distance or density method in the local outlier factor algorithm, and so on. Anomaly types can be *point* or *collective,* and one selects the algorithm for detection based on the anomaly type in a dataset. *Figures 5.7a and 5.7b* show examples of these anomaly types. A point anomaly is a global behavior while a collective anomaly is a local abnormal (non-normal) behavior. There can also be datasets wherein an anomaly can be entirely contextual, which most of the time is visible in time-series data.

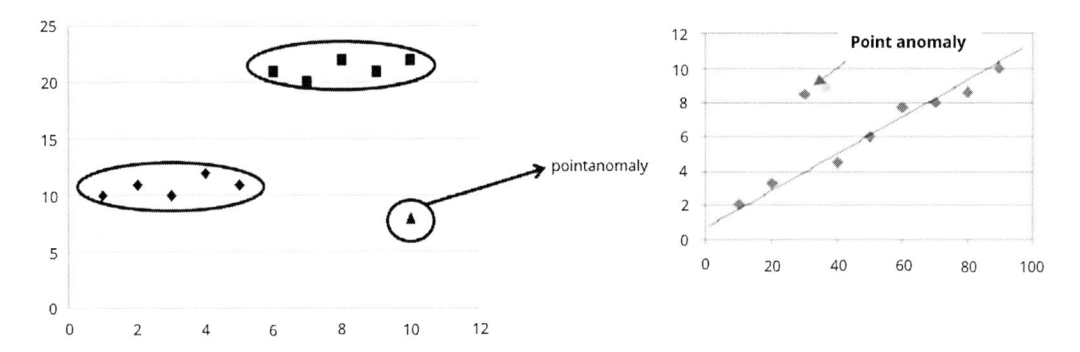

Figure 5.7a: Examples of point anomalies

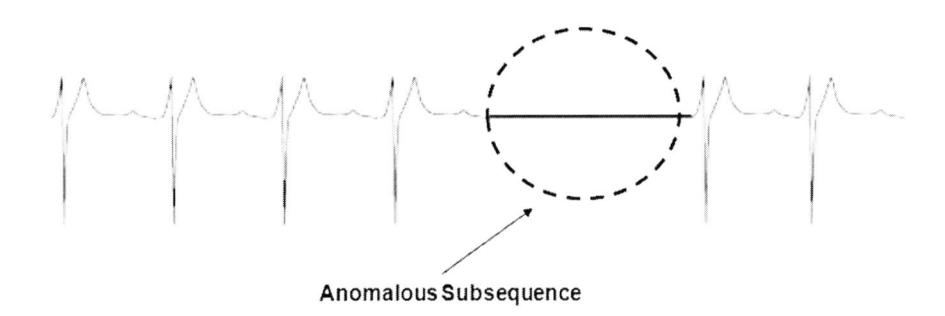

Figure 5.7b: Example of a collective (non-point) anomaly

In the following section, we will implement a one-class SVM solution using Python, as this solution in general has proven to be useful for problems where the (point) outliers forming the minority class lack structure and are predominantly noisy examples (i.e. severe deviations from inliers).

Implementation of SVM

The one-class SVM algorithm does not use (ignores) the examples that are far from or deviated from the observations during training. Only the observations that are most concentrated or dense are leveraged for (unsupervised) learning and such an approach is effective in specific problems where very few deviations from normal are expected.

A synthetic dataset is created to implement SVM. We will have about 2% of the synthetic data in the minority class (outliers) denoted by 1 and 98% in the majority class (inliers) denoted by 0, and leverage the RBF kernel to map the data into a high-dimensional space. The Python code (with the scikit-learn library) runs as follows:

```
import pandas as pd, numpy as np
from collections import Counter
import matplotlib.pyplot as plt
from sklearn.datasets import make_classification
from sklearn.svm import OneClassSVM
from sklearn.model_selection import train_test_split
from sklearn.metrics import classification_report
X, y = make_classification(n_samples = 10000, n_features = 2, n_
informative = 2,
                            n_redundant = 0, n_classes = 2,
                            n_clusters_per_class = 1,
                            weights = [0.98, 0.02], class_sep = 0.5,
random_state = 0)
#Dataset as pandas dataframe
df = pd.DataFrame({'feature1': X[:, 0], 'feature2': X[:, 1], 'target':
y})
#Split dataset into train and test subsets in the ratio 4:1
X_train, X_test, y_train, y_test = train_test_split(X, y, test_size =
0.2, random_state = 42)
#Train SVM model with RBF
one_class_svm = OneClassSVM(nu = 0.01, kernel = 'rbf', gamma =
'auto').fit(X_train)
#nu (specifies number of outliers) = 1% , gamma is a parameter for
nonlinear kernels
prediction = one_class_svm.predict(X_test)
prediction = [1 if i == -1 else 0 for i in prediction] #outliers
denoted by 1, inliers by 0
print(classification_report(y_test, prediction))
```

The report of the classifier (*Figure 5.8*) clearly shows that the one-class SVM model has a recall of 23%, which means the model captures 23% of outliers. The F1-score is the harmonic mean of the two measures, namely precision and recall:

```
              precision    recall  f1-score   support

           0       0.98      1.00      0.99      1957
           1       0.67      0.23      0.34        43

    accuracy                           0.98      2000
   macro avg       0.83      0.62      0.67      2000
weighted avg       0.98      0.98      0.98      2000
```

Figure 5.8: Classification report of one-class SVM

We will visualize the outliers using the following code:

```
#Visualization of outliers
df_test = pd.DataFrame(X_test, columns = ['feature1', 'feature2'])
df_test['y_test'] = y_test
df_test['svm_predictions'] = prediction
fig, (ax1, ax2) = plt.subplots(1, 2, figsize = (16, 8))
ax1.set_title('Original Data')
ax1.scatter(df_test['feature1'], df_test['feature2'], c = df_test['y_test'])
ax2.set_title('One-Class SVM Prediction')
ax2.scatter(df_test['feature1'], df_test['feature2'], c = df_test['svm_predictions'])
```

The default threshold of the algorithm for identifying these 2% outliers can also be customized so that fewer or more data points are labeled as outliers depending on the use case. What is evident from *Figure 5.9* is that most of the outliers (yellow) have been detected correctly by the classifier:

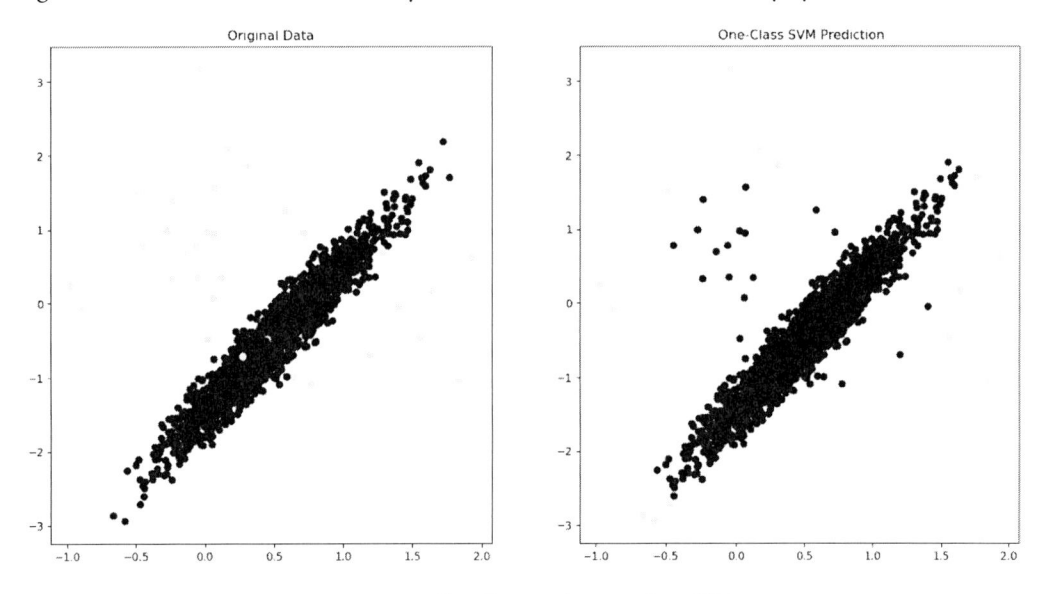

Figure 5.9: Classification by one-class SVM

One-class SVM is particularly useful as an anomaly detector and finds wide usage in sensor data captured from machines in the manufacturing industry.

Summary

In this chapter, we explored SVM as a classifier. In addition to linear data, SVMs can efficiently classify non-linear data using kernel functions. The method used by the SVM algorithm can be extended to solve regression problems. SVM is utilized for novelty detection as well, wherein the training dataset is not polluted with outliers and the algorithm is exploited to detect a new observation as an anomaly, in which case the outlier is called a novelty.

The next chapter is about graph theory, a tool that provides the necessary mathematics to quantify and simplify complex systems. Graph theory is the study of relations (connections or edges) between a set of nodes or individual entities in a dynamic system. It is an integral component of ML and DL because graphs provide a means to represent a business problem as a mathematical programming task in the form of nodes and edges.

6
Graph Theory

Graphs are mathematical structures that are used to model pairwise relationships. Graph theory provides a tool to quantify these relationships in a dynamic system. In other words, graphs are ways to represent a network or a collection of interconnected objects. Graph theory is mostly applied in operations research and social sciences. Its history dates back to the 18th century when the Swiss mathematician Leonhard Euler solved the Königsberg bridge problem (*Figure 6.1*), which served as a precursor to graph theory. The city of Königsberg in Russia was set on both sides of the Pregel river and included two large islands, namely, Kneiphof and Lomse, which were connected by seven bridges. The problem was to devise a walk through the city that would cross each of these bridges only once. Euler drew out the first known visual representation of a modern graph for the city. It is represented (abstractly) by a set of points known as vertices or nodes, connected by a set of lines known as edges. Edges represent the relationships between nodes.

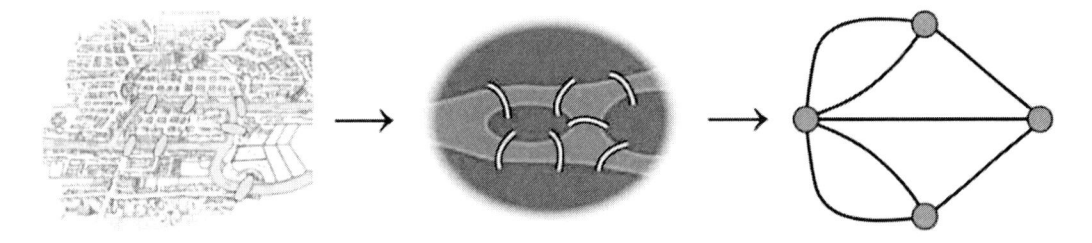

Figure 6.1: Seven bridges of Königsberg

A graph theory problem typically uses the framework of mathematical optimization, which has three components, namely, the objective function, decision variables, and constraints. The objective function is minimized to obtain the optimal path between nodes in a graph because there can be multiple paths connecting two nodes (multigraphs). The objective of using a graph to solve a problem is multifold. One could be to visualize the edges, figure out closely connected nodes, and identify the nodes that directly influence the objective function. A possible decision variable among others could be whether or not to add an edge between two nodes. A typical constraint could be the degree each node can attain, that is, the maximum number of connections each node can have to other nodes in the network. A well-known problem solved using graph theory is the **traveling salesman problem** (**TSP**), in which the shortest path starts and ends at the same vertex/node and visits each edge exactly once. Such examples (routing) are related to the field of linear programming.

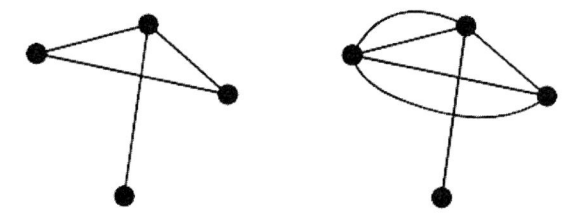

Figure 6.2a: Simple graph (L) and multigraph (R)

Graphs come in a variety of sorts, the most common of which are simple graphs and multigraphs. These are shown in *Figure 6.2a* and there is also a graph or directed graph (edges have directions) together with a function that assigns a positive real number to each edge, known as a network. A network is in fact an oriented edge-labeled graph, as shown in *Figure 6.2b*.

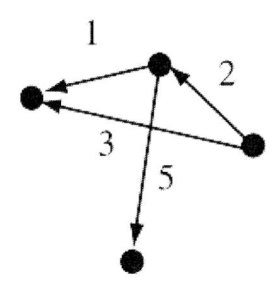

Figure 6.2b: Graph as network (model)

There is a restricted type of graph known as trees. Tree data structures are different from graph data structures (*Figure 6.3*). A tree, which is a hierarchical model, can never have cyclical links (edges) like some graphs have. Trees are **directed acyclic graphs** (**DAGs**) and are unidirectional. A graph has no root (source) node, whereas a tree does have this node, along with child nodes, and every child has one parent node.

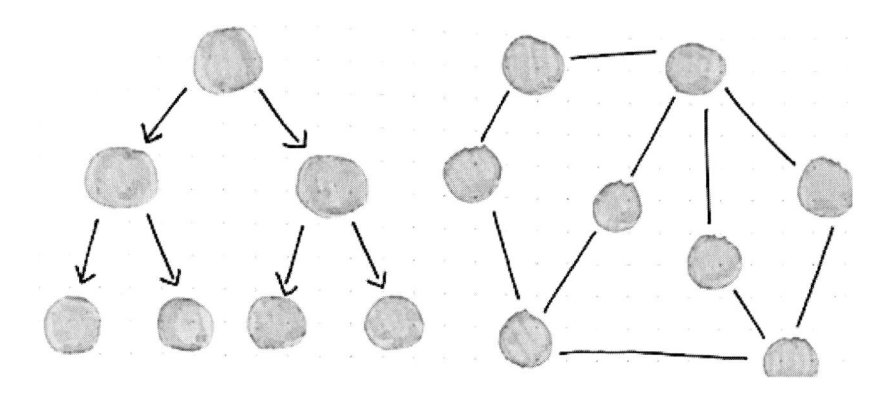

Figure 6.3: Tree (L) with n-1 edges (n = number of nodes) and graph
(R) with no rule about the number of edges it can have

Graphs can also have loops, circuits, and self-loops. There are databases that use graph structures for semantic queries with nodes and edges. Querying relationships (edges labeled, directed) is fast in general, as they are perpetually stored in the database. Graph databases are commonly called NoSQL. Graphs are utilized in analyzing social networks as well.

This chapter covers the following topics:

- Types of graphs
- Optimization use case
- Graph neural networks

The next section discusses the types of graphs depending on labels, directions, and the weights of edges in the graphs.

Types of graphs

The primary graph types are undirected, directed, and weighted graphs, as illustrated in *Figure 6.4*. Social networks can be undirected as well as directed graphs. In the former, edges end up being unordered pairs, for example, Facebook. In the latter, edges are ordered pairs, for example, Twitter, in which one node is an origin and the other a destination.

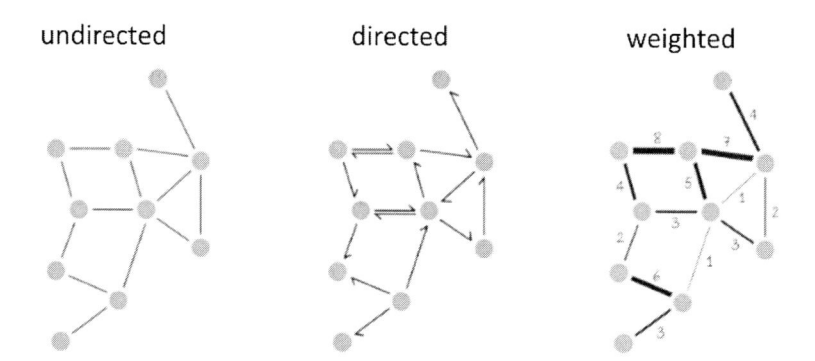

Figure 6.4: Three standard graph types

We will explore each of these graphs in the following subsections.

Undirected graphs

While solving a problem using graph theory, the first step is to determine the type of graph we are dealing with. In undirected graphs, there is no particular direction of the edges between nodes, in other words, the edge is bidirectional. An edge connecting node 1 to node 2 (*Figure 6.5*) would be identical to the edge connecting node 2 to node 1.

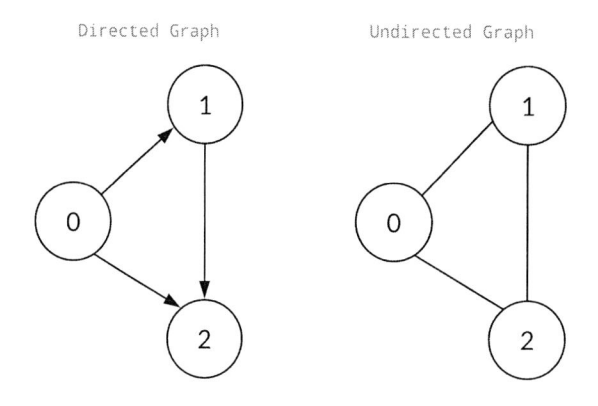

Figure 6.5: Directed and undirected graphs

Directed graphs

In directed graphs, or digraphs, there is a specified direction between the nodes. The edge between nodes 1 and 2 is directed from 1 toward 2 (*Figure 6.5*) and a link directed toward 1 from 2 would not be permitted. In other words, the edges between nodes are unidirectional.

Weighted graphs

If the edge between two nodes or vertices has an associated weight to represent implications such as distance or cost, the corresponding graph is said to be weighted. Weighted graphs can be either directed or undirected (*Figure 6.6*). Weighted graphs are applicable to many real-world scenarios, for example, search engines comparing flight times and cost or route planning.

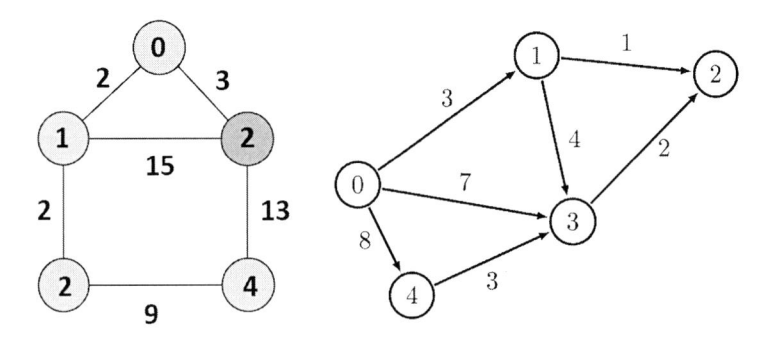

Figure 6.6: Undirected weighted graph (L) and directed weighted graph (R)

Any graph can be represented mathematically through an adjacency matrix, which describes all permitted routes or paths between nodes in the graph. An adjacency matrix for a directed graph is shown in *Figure 6.7a*, and that of an undirected graph is illustrated in *Figure 6.7b*.

Adjacency Matrix

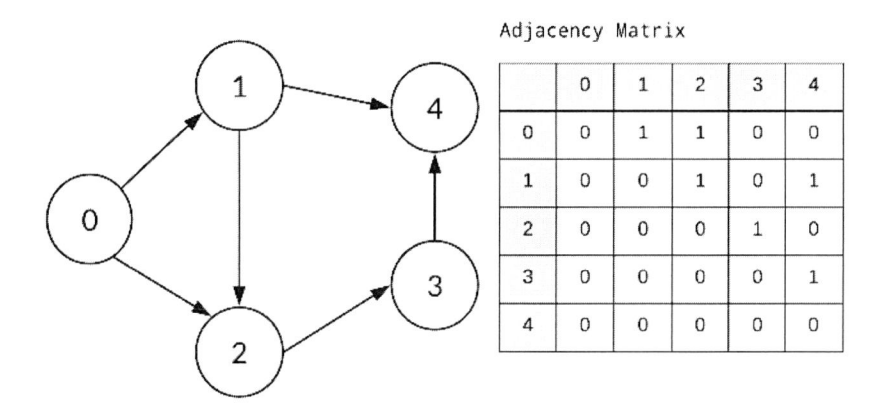

	0	1	2	3	4
0	0	1	1	0	0
1	0	0	1	0	1
2	0	0	0	1	0
3	0	0	0	0	1
4	0	0	0	0	0

Figure 6.7a: Adjacency matrix of a directed graph

We know by now that graphs make networks more interpretable and easier to visualize. More computer memory is consumed by an adjacency matrix for a bigger graph (more nodes) like the one of the weighted graph shown in *Figure 6.7c*.

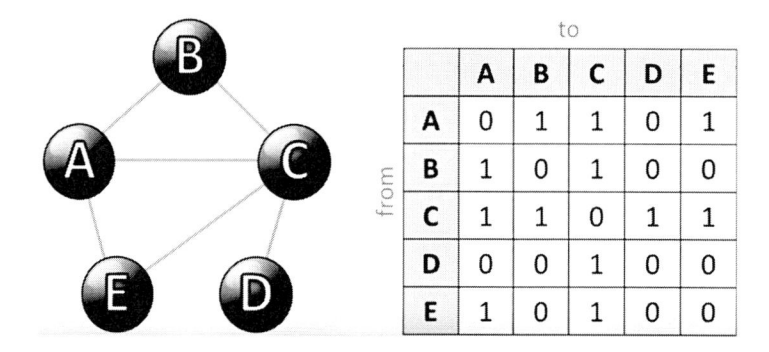

Figure 6.7b: Adjacency matrix of an undirected graph

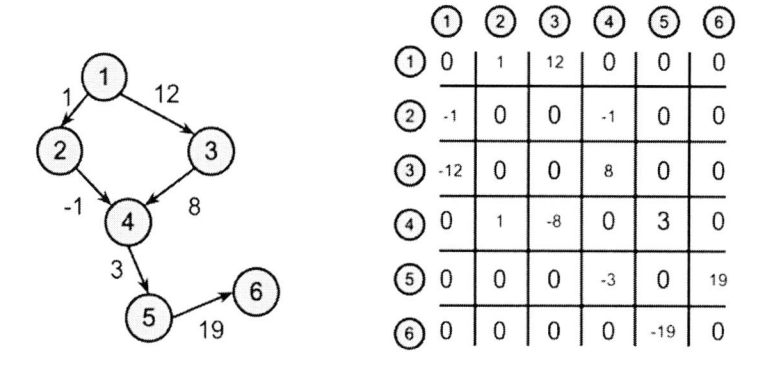

Figure 6.7c: Adjacency matrix of a weighted (directed) graph

Most adjacency matrices are sparse; that is, the graphs are not densely connected, making computations harder.

Now that we have explored the different graphs, we will investigate a use case of an optimization problem solved using graph theory.

Optimization use case

Graphs can be used to model relations and processes in physical, biological, and information systems. They have a wide range of applications, such as ranking hyperlinks in search engines, the study of biomolecules, computer network security, GPS in maps to find the shortest route, and social network analysis. There are knowledge graphs for information mining as well. In the following subsection, we pick a dataset and formulate the problem in a way that is solved using graph theory.

Optimization problem

There can be multiple paths between origin and destination airports. An airline seeks the shortest possible path between airports, wherein the shortest path can be defined in terms of either distance or airtime. If the city airports are represented as nodes and the flight routes between them as edges, we convert the problem into a graph (*Figure 6.8a*). The dataset can be found in the GitHub repository: `https://github.com/ranja-sarkar/graphs`.

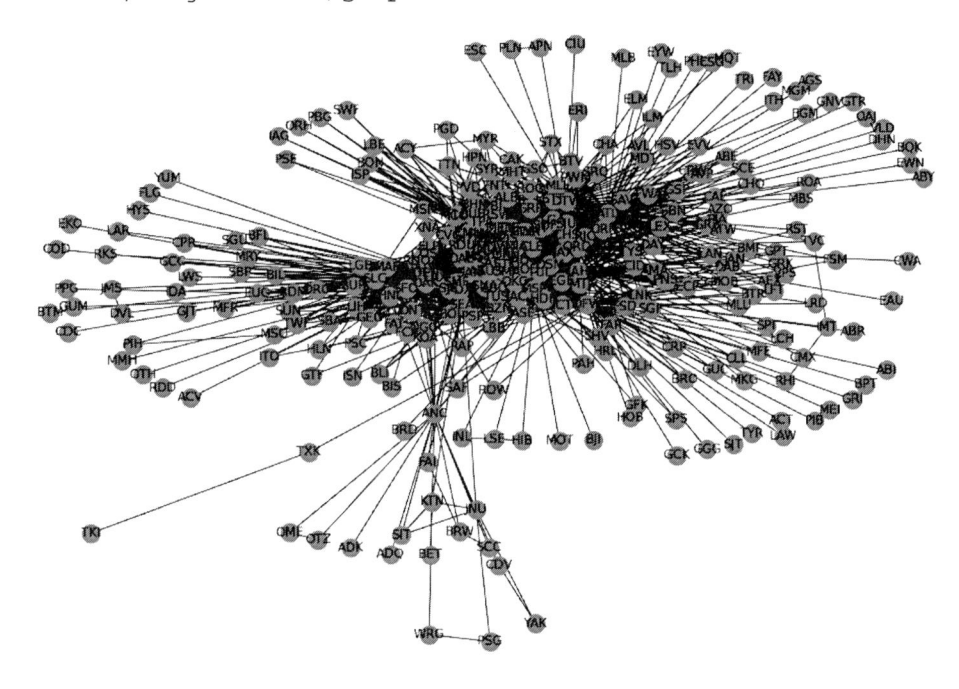

Figure 6.8a: Network (flight routes) between origin (city) airport and destination (city) airport

We can identify the shortest (minimum airtime or minimum distance) possible path between any two city airports from the graph. Example code in Python to arrive at the solution is explained in the following subsection.

Optimized solution

The dataset, a sample of which is displayed in *Figure 6.8b*, has records of flights in January 2017 from the USA out of the origin (source) city, which is given by `Origin`, to the destination city, which is given by `Dest`. The distance between the origin and destination and the airtime of the flight are the most relevant variables required to find the optimized solution.

68   Graph Theory

Year	DayOfWeek	FlightDate	Origin	OriginCityName	Dest	DestCityName	CRSDepTime	DepTime	CRSArrTime	ArrTime	AirTime	Distance	DistanceGroup
2017	2	17-01-2017	CLT	Charlotte, NC	PHX	Phoenix, AZ	1619	1616	1856	1842	244	1773	8
2017	3	18-01-2017	CLT	Charlotte, NC	PHX	Phoenix, AZ	1619	1614	1856	1821	228	1773	8
2017	4	19-01-2017	CLT	Charlotte, NC	PHX	Phoenix, AZ	1619	1611	1856	1826	236	1773	8
2017	5	20-01-2017	CLT	Charlotte, NC	PHX	Phoenix, AZ	1619	1656	1856	1929	252	1773	8
2017	6	21-01-2017	CLT	Charlotte, NC	PHX	Phoenix, AZ	1619	1632	1856	1858	245	1773	8
2017	7	22-01-2017	CLT	Charlotte, NC	PHX	Phoenix, AZ	1619	1636	1856	1921	254	1773	8
2017	1	23-01-2017	CLT	Charlotte, NC	PHX	Phoenix, AZ	1619	1616	1856	1907	264	1773	8
2017	2	24-01-2017	CLT	Charlotte, NC	PHX	Phoenix, AZ	1619	1619	1856	1904	261	1773	8
2017	3	25-01-2017	CLT	Charlotte, NC	PHX	Phoenix, AZ	1619	1616	1856	1906	262	1773	8

Figure 6.8b: Dataset for the case study

You can have a look at the nodes and edges of the corresponding graph resulting from executing the code. The visualization in *Figure 6.7a* represents the network of flights:

```
import pandas as pd, numpy as np
import networkx as nx
import matplotlib.pyplot as plt
#dataset
data = pd.read_csv('/kaggle/input/airline/FlightsUSA.csv')
df = nx.from_pandas_edgelist(data, source = 'Origin', target = 'Dest',
edge_attr = True)
#df.nodes()
#df.edges()
plt.figure(figsize = (18,12))
nx.draw_networkx(df, with_labels = True)
#plot the graph
```

A passenger wishing to take the shortest route from LAS (Las Vegas) to PBI (Palm Beach in Florida) with respect to the distance and airtime metrics can run the piece of code (consuming Dijkstra's shortest-path algorithm contained in the NetworkX Python library) and self-serve or decide on the best route:

```
shortest_airtime = nx.dijkstra_path(df, source = 'LAS', target =
'PBI', weight = 'AirTime')
shortest_dist = nx.dijkstra_path(df, source = 'LAS', target = 'PBI',
weight = 'Distance')
print(shortest_dist,shortest_airtime)
```

The shortest path between LAS and PBI based on distance is shown in the output in *Figure 6.9a*.

$$['LAS', 'DFW', 'PBI']$$

Figure 6.9a: Output when the model parameter is distance

The shortest path based on airtime is shown in *Figure 6.9b*.

$$['LAS', 'IAH', 'PBI']$$

Figure 6.9b: Output when the model parameter is airtime

The algorithm creates the shortest path set first by picking the vertex closest to the source vertex, then a vertex from the remaining ones closest to the source. This continues until the set includes all vertices/nodes. Graph theory can therefore be applied to travel planning and finding the best route to deliver post, among other usages. The mathematical formulation of graphs is intuitive and comprehensive. In the next section, we will introduce **graph neural networks** (**GNNs**), which will involve diving into **deep learning** (**DL**).

Graph neural networks

DL algorithms make use of graphs to predict at the level of nodes, edges, or entire graphs. In node classification, the label of samples (nodes) is determined by looking at the labels of neighbors. In graph classification, the entire graph is classified into different categories, an example being categorizing documents using natural language processing. The relationships (edges) between nodes or entities are utilized in recommendation systems. Image and text are types of structured data that can be described as grids of pixels and sequences of words, respectively. These are shown in *Figure 6.10a*. Graphs, in contrast, are unstructured data. Graphs can contain any kind of data, including images and text.

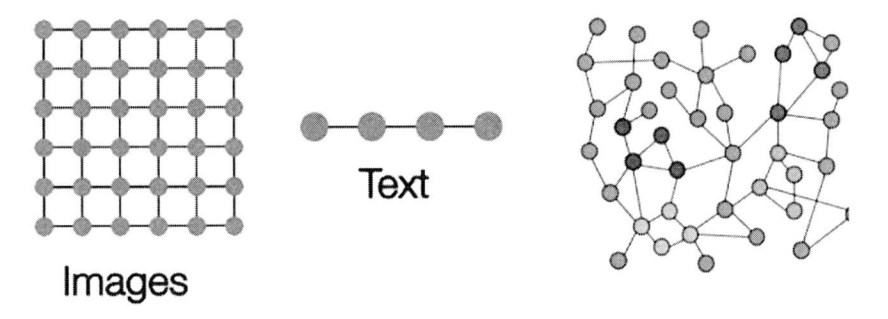

Figure 6.10a: Structured data (L) as opposed to graphs/networks (R)

GNNs organize graphs using a process called message passing so that DL algorithms can use the embedded information about the neighbors of each node to find patterns and make predictions. Typically, the input to a GNN pipeline (*Figure 6.10b*) is a defined graph structure with its type and scale.

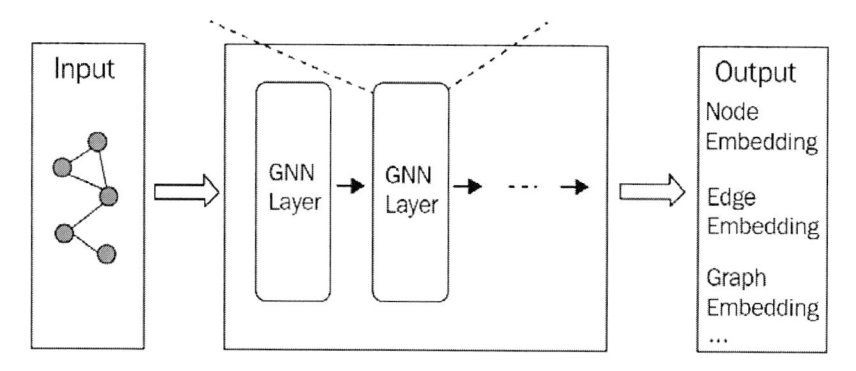

Figure 6.10b: GNN has a graph as an input

In graph theory, the concept of node embedding is implemented, meaning mapping nodes to a lower dimensional (than the actual dimension) space so that similar nodes in the graph are close to each other. *Figure 6.11* illustrates how information from the input graph will propagate to the outside of the neural networks (gray boxes). Therefore, the aggregation of information takes multiple (three here) layers. A model can be trained by supervised and unsupervised means. In the latter, only the graph structure is used and similar nodes have similar embeddings. The former is used for a supervised task such as node classification.

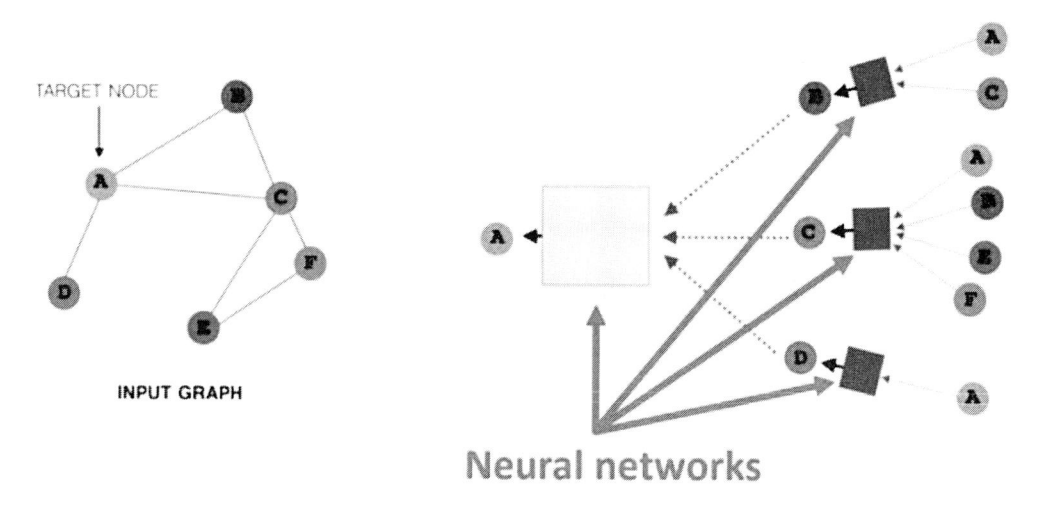

Figure 6.11: Propagation of information

GNN architectures are used in image and text classification problems. They are also used in relation (semantic) extraction. They have become powerful tools in recent years for any problem that can be modeled by graphs.

Summary

In this chapter, we learned about a theory that is helpful in simplifying and quantifying complex connected systems called networks. Graph theory is the study of relationships (represented as edges in graphs) between dynamic entities and helps better interpret network models. We further elaborated (with Python code) on how an optimization problem can be mathematically formulated and solved using this concept. A lot of problems can be approached using a graph framework that involves the components of mathematical optimization, as discussed in a section of this chapter.

This chapter also introduced GNNs, which operate on the structure and property of a graph. A single property is predicted for an entire graph for a graph-level task, a property of each node is predicted for a node-level task, and the property of each existing edge in a graph is predicted abstractly an edge-level task. GNNs are applied when graphs are complex and deep.

In the next chapter, we will study the Kalman filter, which is one of the most efficient estimation algorithms. It provides a recursive computation method to estimate the (unknown) state of a discrete data-controlled process by using a series of measurements that are typically noisy, as well as calculating the uncertainty in measurement. Kalman filtering is a concept applied to topics such as signal processing, wherein the variables of interest can be indirectly measured as they are unable to be directly measured.

7
Kalman Filter

In a dynamic system, there is uncertain information. To capture the uncertainty, yet another mathematical tool, called the Kalman filter, comes into play. One can utilize the Kalman filter to optimally estimate the system's next state, and it is ideal for continuously changing systems. It is especially useful for handling noisy sensor data by collating sensor data to best estimate the parameter of interest. In other words, the Kalman filter is an estimator of the system's states in the presence of imprecise and uncertain measurements. It is mostly useful for the estimation of unobserved variables in real time.

The Kalman filter algorithm is widely used in signal processing, target tracking, navigation, and control applications. In tracking and control systems, an accurate and precise estimation of location and velocity, which are hidden (unknown) states, is a challenge. The uncertainty in the measurement of hidden states is attributed to external factors, such as atmospheric effects and thermal noise. The Kalman filter is an algorithm to estimate the hidden states of a dynamic system and predict the future state of the system based on past estimations. It is named after Rudolf E. Kalman, who published his famous paper on a recursive solution to a discrete data linear-filtering problem in 1960.

A system is governed by a set of equations, and this set is called a dynamic model or state space model. If a system's current state and the dynamic model are known, the subsequent state can be estimated. The uncertainty in the dynamic model is influenced by external factors and is called process noise. This is the error or misalignment between the equations of motion and the actual motion of the system. The random error or uncertainty in measurement is called measurement noise. In order to improve the estimation of the future state of the system, it becomes essential to account for process noise as well as measurement noise. The Kalman filter takes both of these uncertainties into account.

Any measured (or computed) parameter is an estimate and it can be significantly improved by the usage of multiple sensors. In this regard, two terms, namely, accuracy and precision, have to be understood well as they cannot be used interchangeably. Accuracy indicates the closeness of the measurement to the true value, while precision indicates the variability in measurements of the same parameter. Accuracy and precision form the basis of an estimate of a hidden state of the system. *Figure 7.1* shows the high accuracy and high precision of an estimate:

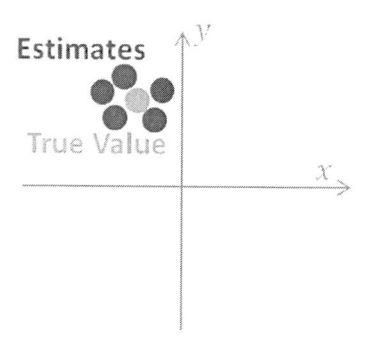

Figure 7.1: High accuracy and high precision of measurement

Unbiased systems have no or significantly low built-in systematic error (bias) and, hence, are high-accuracy systems. Real systems are biased and have process noise. High-precision systems have low variance (or low uncertainty). The influence of variance can be reduced by averaging (smoothing) measurements. The more measurements, the closer the estimate to the true value. A **probability distribution function** (**PDF**) describes a measurement as it is a random variable. The dispersion in distribution (*Figure 7.2*) shows the measurement noise. In a low-precision and low-accuracy system, the estimates are neither close to each other nor to the true value. They will be spread all over the four quadrants of spatial coordinates.

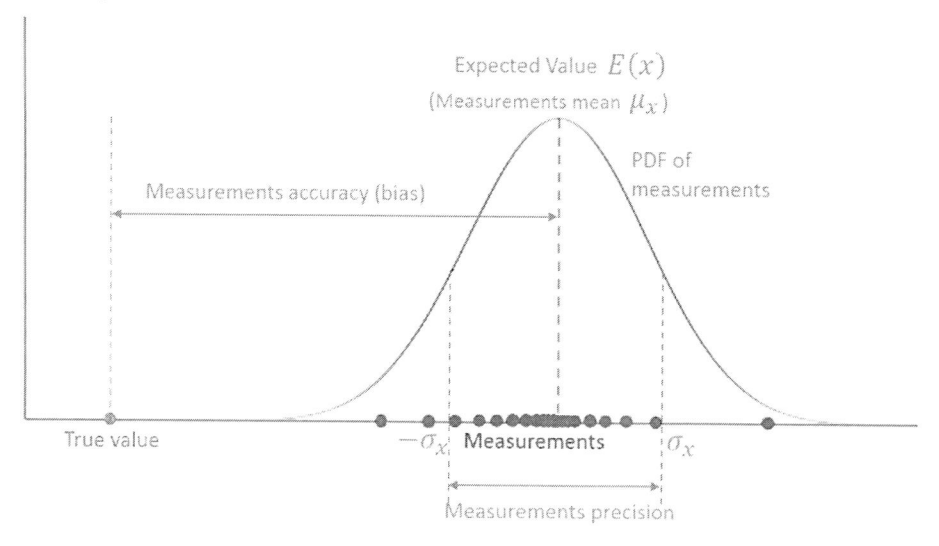

Figure 7.2: Measurement distribution

Now that the concepts of a measurement and its precision and accuracy are clear, we will discuss how the Kalman filter works. This chapter covers topics that are essentially the component steps of this estimation algorithm, concluding with some illustrative Python code:

- Computation of measurements
- Filtration of measurements
- Implementation of the Kalman filter

The Kalman filter will be tested out with an example to estimate the position (displacement) and velocity of a moving object using Python in the last section.

Computation of measurements

We will start with a flow diagram of the Kalman filter algorithm, shown in *Figure 7.3a*. The Kalman filter requires an initial guess to start with. This input can be a very rough estimate (zeroth step). So, step 0 is the initial guess and step 1 is the measurement of the state variable.

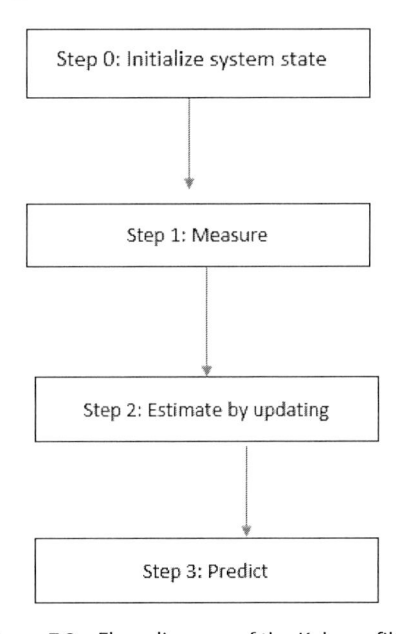

Figure 7.3a: Flow diagram of the Kalman filter

When the input is a measured value, the output is the current state estimated using the state update equation in step 2, which is calculated from the predicted value of the current state and the residual scaled (updated) by a factor called the Kalman gain. The Kalman gain takes the input measurement uncertainty into account, the residual being the difference between the measured and predicted values. This update and estimate make the second step in the algorithm.

The output from step 2 is fed to predict the next state of the system. The state for the next iteration is predicted using the dynamic model. The prediction in step 3 is basically an extrapolation of the current state utilizing the set of equations of dynamism. This continues for several iterations and the Kalman gain is calculated at each iteration. An illustrative example of tracking the values (true, measured, estimated, and predicted) of a constant velocity aircraft in one dimension is shown in *Figure 7.3b*, which also exemplifies a univariate Kalman filter:

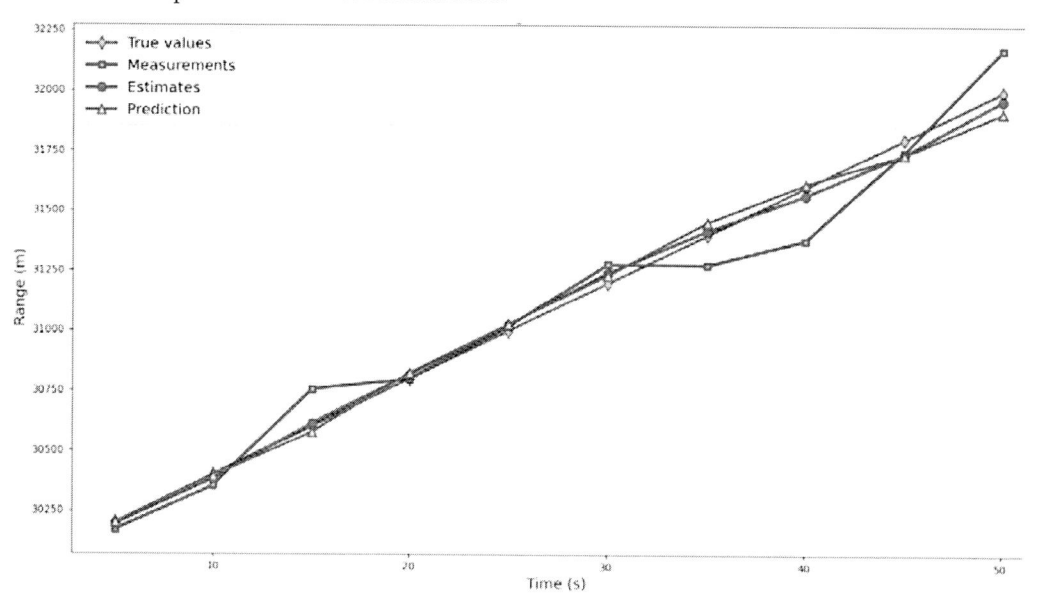

Figure 7.3b: Estimated and predicted values (positions) of constant velocity aircraft

It is evident from the figure that the estimation algorithm (Kalman filter) has a smoothing effect on the measurements and converges toward the true value with an increasing number of iterative steps. The next section describes the filtration of random variables in the form of measurements to optimize the uncertainty in the estimate.

Filtration of measurements

The Kalman filter has inputs and outputs like any filter. The inputs are noisy and inaccurate measurements, while the outputs are much less noisy and more accurate estimates. Mathematically speaking, the inputs to the filter are a measured value and measurement covariance matrix. The dynamic system model is the state transition matrix (representing equations) and process noise covariance matrix, the Kalman gain is internal and dependent on the system, and the outputs from the filter are the state variable and state covariance matrix. This is illustrated in *Figure 7.4a*:

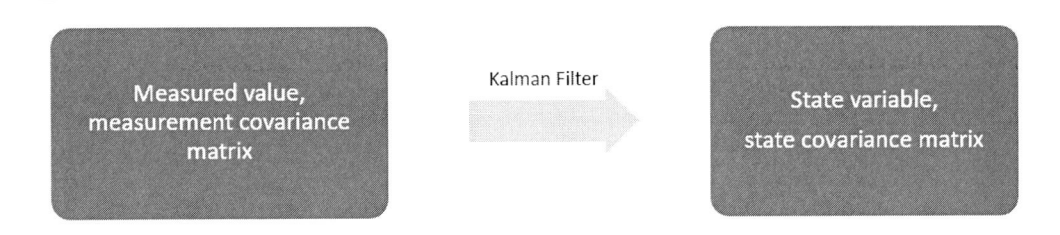

Figure 7.4a: Input and output of the Kalman filter

When an estimate is propagated in time, the future state is inherently uncertain and hence, the error covariance matrix grows with time. The dynamic model (equations of motion) is approximate; the process noise (uncertainty) adds to the existing noise, and this is represented by the process noise covariance matrix. The estimate needs to be converted from the state space to the measurement space and this conversion is done via another matrix (state-to-measurement). For Kalman filters applied to linear systems discussed in this chapter, this transformation matrix is simple, whereas the transformation can be complex for non-linear (or extended) Kalman filters. If the system is non-linear, a non-linear state estimator or filter is utilized. For example, extended Kalman filters do a linearization of the distribution around the mean of the current estimate and use it in the predict and update states of the algorithm.

The Kalman gain is computed in each iteration and it determines the influence of the input measurement (new information) on the estimate. If the input measurement is very noisy, the Kalman gain will trust its current state estimate more than the input. The Kalman filter has the ability to recognize how to appropriately put weights on its current estimate and a new input measurement at each time step to produce an optimal estimate.

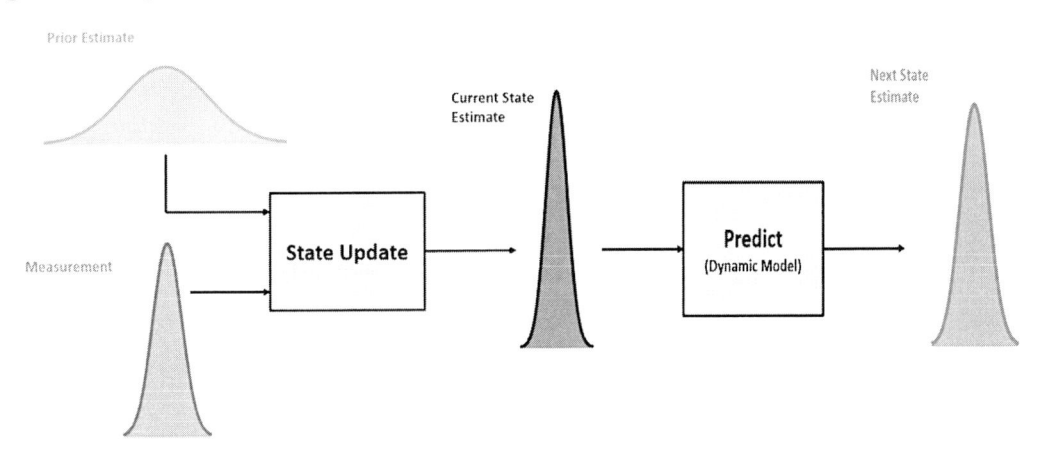

Figure 7.4b: Kalman filtering in the form of distribution functions

To summarize, the Kalman filter is an optimal filter that treats two random variables (prior estimate and measurement) to estimate the current state variable, as shown in *Figure 7.4b*, by minimizing the uncertainty. If we have a prior distribution with a high variance and a measured distribution with a lower variance, the Kalman filter combines the two to estimate a distribution with a higher peak and narrower variance than the prior.

In the following section, the computation and filtration of measurements leveraging the Kalman filter estimation algorithm are implemented with Python code.

Implementation of the Kalman filter

In this illustrative example, time-series data is used as input and the Kalman filter provides estimates at each time step. The example is of a moving vehicle for which initializations of velocity, displacement, and acceleration are made. Acceleration values at different time steps are also incorporated. The kinematic equation, which relates displacement, velocity, and acceleration of the vehicle, yields the true values:

```
import numpy as np
import math, random
import matplotlib.pyplot as plt

current_vel, current_disp, current_accel = 2, 0, 0
total_time = 100
accel_dict = {0:0,5:2,10:8, 20: -2,40:5,45: 9, 60: -3,85:0}
true_values = []
for t in range (1, total_time+1):
    current_disp = current_disp + current_vel + (1/2) * current_accel
    try:
        current_accel = accel_dict[t]
    except KeyError:
                pass
    current_vel = current_vel + current_accel
    true_values.append((current_disp, current_vel, current_accel))
```

Random noise (error) is added to the true values in the form of small perturbations, and measurements are determined:

```
err_range = [700, 30, 15] #noise
measurements = []
for item in true_values:
    d,v,a = item
    random_err = [random.randint(-1*err_range[0], err_range[0]),
random.randint(-1*err_range[1], err_range[1]), random.randint(-1*err_range[2], err_range[2])]
    new_disp = d + random_err[0] if d+random_err[0] > 0 else 0
```

```
new_vel = v + random_err[1]
new_accel = a + random_err[2]
measurements.append((new_disp, new_vel, new_accel))
```

We can compare the true values with the measured values of displacement. On running the following piece of code, we obtain a visual comparison (*Figure 7.5*). Similarly, we can make a visual comparison of true values with the measurements of velocity as well:

```
plt.plot([i for i in range(total_time)], [y[0] for y in true_values],
'r--', label = 'True Values')
plt.plot([i for i in range(total_time)], [y[0] for y in measurements],
'b--', label = 'Measurements')
plt.ylabel("Displacement")
plt.xlabel("Time (s)")
plt.legend()
plt.show()
```

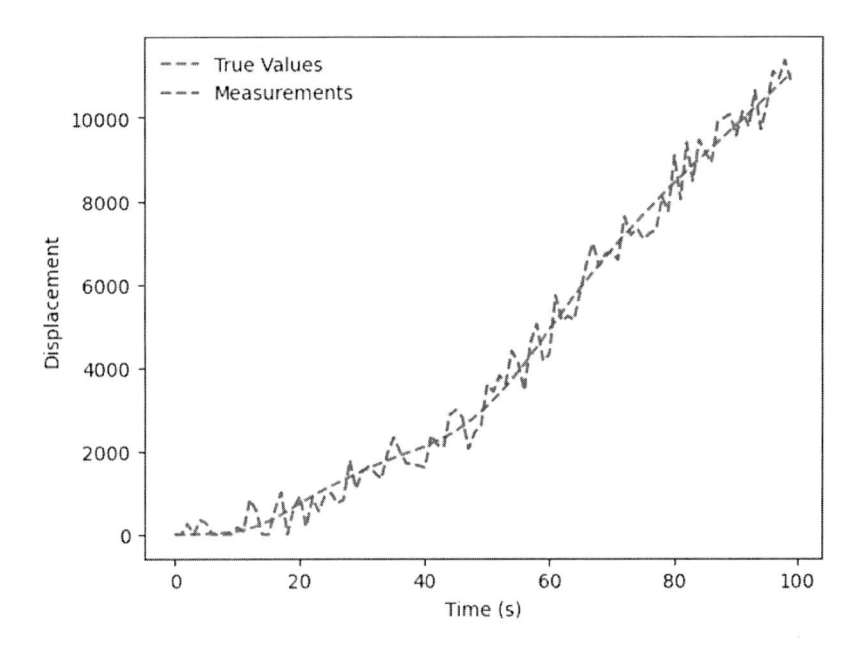

Figure 7.5: True values versus measurements for displacement

We are, therefore, done with the computation of measurements. Next is the filtration of measurements for which the noisy data is fed into the Kalman filter. Displacement and velocity are initialized and the error covariance (Q) is estimated. The transition matrix (A) is also used wherein it is assumed acceleration of the vehicle is unknown. The measurement error (R) is higher than the estimation error due to noise in the data. H in the following code yields states and P is the error matrix:

```
x_k = np.asarray([30,20])
Q = np.asarray([[0.004,0.002],[0.002,0.001]])
A = np.asarray([[1,1],[0,1]])
R = np.asarray([[0.4,0.01],[0.04,0.01]])
H = np.asarray([[1,0],[0,1]])
P = np.asarray([[0,0],[0,0]])
estimation = []
for k_loop in range(total_time):
      z_k = np.asarray([measurements[k_loop][0], measurements[k_loop]
[1]])
      x_k = A.dot(x_k)
      P = (A.dot(P)).dot(A.T) + Q
            K = (P.dot(H.T)).dot(np.linalg.inv((H.dot(P).dot(H.T)) +
R))
      x_k = x_k + K.dot((z_k - H.dot(x_k)))

      P = (np.identity(2) - K.dot(H)).dot(P)
      estimation.append((x_k[0], x_k[1]))
```

We can now compare the true values with an estimation of displacement as well as velocity. On running the following piece of code, we obtain a visual comparison (*Figure 7.6*) of displacement:

```
plt.plot([i for i in range(total_time)], [y[0] for y in true_values],
'r--', label = 'True Values')
plt.plot([i for i in range(total_time)], [y[0] for y in measurements],
'b--', label = 'Measurements')
plt.plot([i for i in range(total_time)], [y[0] for y in estimation],
'g--', label = 'Estimated Values')
plt.title('Estimation of displacement')
plt.ylabel("Displacement")
plt.xlabel("Time (s)")
plt.legend()
plt.show()
```

Estimation of displacement

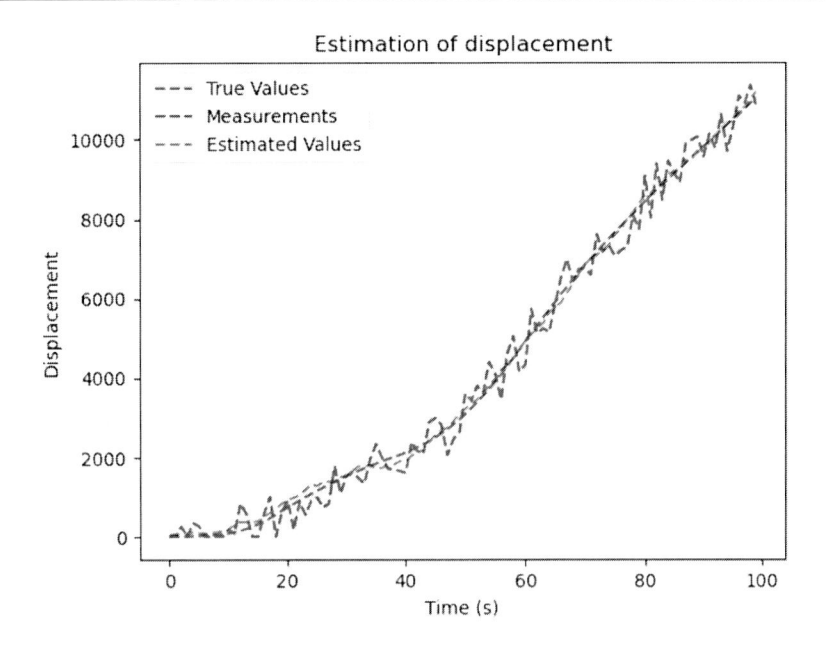

Figure 7.6: True values versus estimated values of displacement

It is evident that the Kalman filter provides estimates of the displacement of a moving vehicle that are very close to the true values when used on noisy data. The example of a moving vehicle used in the algorithm implementation is illustrated in *Figure 7.7*:

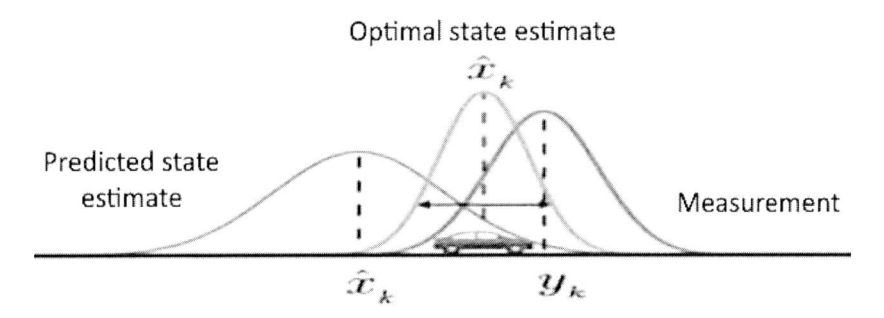

Figure 7.7: Optimal state estimate of a moving vehicle

Though synthetic data was utilized in the previous code, the usage and application are generic, and the algorithm works for a dynamic system provided the matrices governing the system dynamics are set up properly.

Summary

In this chapter, we explored the Kalman filter – the estimation and prediction algorithm utilized to solve problems in signal processing, navigation, and control systems. There are linear and univariate (one-dimensional) Kalman filters in which the system dynamics are assumed to be linear. Many dynamic processes, however, have more than one dimension, and in such cases, we utilize multivariate and mostly non-linear (or extended) Kalman filters. For example, the state vector that describes a moving object's position and velocity in space is six-dimensional, and a non-linear Kalman filter is utilized to determine the displacement (and velocity) in space of such an object. Also, the Kalman filter consumes low computational power (leading to a shorter runtime) due to the usage of matrices in its operation that occupy less computer memory. The Kalman filter is arguably the best estimation algorithm with noisy data as it mitigates the uncertainty by combining the information we have and providing us with a distribution we can feel more confident about.

The next chapter is the last one in this part (*Mathematical Tools*) of the book and is about the Markov chain, an algorithm to sample states from a population with a complex probability distribution. It is a probabilistic tool to traverse a system of states. In other words, it randomly walks across a graph and helps predict the next state just from knowledge of the present.

8
Markov Chain

The Markov chain is one of the most important stochastic processes and solves real-world problems with probabilities. A Markov chain is a model of random movement in a discrete set of possible locations (states), in other words, a model of transition from one location (state) to another with a certain probability. It is named after Andrey Markov, the Russian mathematician who is best known for his work on stochastic processes. It is a mathematical system describing a sequence of events in which the probability of each event depends only on the previous event.

"The future depends only upon the present, not upon the past."

The events or states can be written as $\{X_0, X_1, X_2, ...\}$, where X_t is the state at time t. The process {} has a property, which is X_{t+1}, which depends only on X_t and does not depend on $\{X_0, X_1, ..., X_{t-1}\}$. Such a process is called a Markovian or Markov chain. It is a random walk to traverse a system of states. A two-state Markov chain is one in which a state can transition onto itself (that is, staying in the same state). It is shown in *Figure 8.1* (which is a state diagram). An example of a Markov chain is the PageRank algorithm, which is used by Google to determine the order of results for a search.

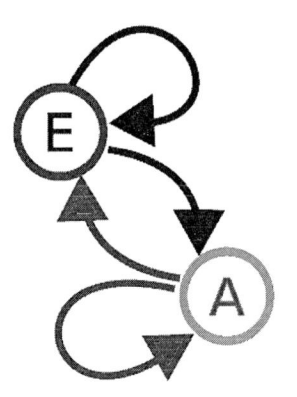

Figure 8.1: Two-state (A and E) Markov chain

Markov chains are quite powerful when it comes to including real-world phenomena in computer simulations. It is a class of probabilistic graphical models representing a dynamic process, the limitation being that it can only take on a finite number of states. Markov chains have no long-term memory (are memory-less, in short) and hence know no past states. Therefore, the only state determining the future state in a Markov chain is the present, and this is called a Markov property.

This chapter covers the following topics:

- Discrete-time Markov chain
- **Markov Chain Monte Carlo (MCMC)**

The following section discusses the very foundation of a Markov chain, which is a discrete-time stochastic process.

Discrete-time Markov chain

For a discrete-time Markov process, $(X_n)_{n\geq 0}$ while in continuous time n is replaced by t where t runs until infinity. Given the present state, past and future states are independent in a Markov chain, which in turn means that the future is only dependent on the present. In the following subsections, we will learn about the transition matrix and an application of the Markov chain in time-series data for short-term forecasting.

Transition probability

The transition probabilities between Markov states are captured in a state transition matrix. The dimension of the transition matrix is determined by the number of states in the state space. Every state is included as a row and a column, and each cell in the matrix gives the probability of transition from its row's state to its column's state, as shown in *Figure 8.2*. In order to forecast one step ahead, one must know the transition matrix and the current state. The transition probability (matrix element) is typically established from historical sequential data.

	E	A
E	P(E|E) = 0.5	P(A|E) = 0.5
A	P(E|A) = 0.5	P(A|A) = 0.5

Figure 8.2: Transition matrix for the two states

Application of the Markov chain

Markov chains model the behavior of a random process. They can be used for text prediction in order to autocomplete sentences or to model the evolution of time-series data, for example, modeling the behavior of financial markets.

An example of modeling the price of stock using a Markov chain is depicted in the following Python code. A set of states (in the order `increase`, `decrease`, and `stable`) is defined for the time evolution of the stock price with the probability of transition between these states. The transition matrix is used to predict the probable future (next state) price:

```
import numpy as np
states = ["increase", "decrease", "stable"] #Markov states
transition_probs = np.array([[0.6, 0.3, 0.1], [0.4, 0.4, 0.2], [0.5, 0.3, 0.2]])
num_steps = 10                   #time-steps for simulation
def MC_states(current_state):
    future_states = []
for i in range(num_steps):
        probs = transition_probs[states.index(current_state)]
        new_state = np.random.choice(states, p = probs)
        future_states.append(new_state)
        current_state = new_state #Update current state
    return future_states
#output
MC_states("increase")
```

The output is a sequence of future states, shown in *Figure 8.3*, given a current state. A different output is obtained if the current state is set to `decrease` or `stable` (initial state) while executing the function in the code. The sequence of states depicts the evolution of the stock price over time. Caution must be exercised when the system does not exhibit stationary behavior, that is, the transition probabilities between states change over time. In that case, a complex Markov model or a different model altogether may be used to capture the system's behavior.

```
['increase',
 'increase',
 'stable',
 'increase',
 'increase',
 'increase',
 'decrease',
 'stable',
 'increase',
 'increase']
```

Figure 8.3: Output of the example code in Python

If x_i is the number of times the sequence is in state i (state is observed) and x_{ij} is the number of times there is a transition from state i to state j, then the transition probability is defined as follows:

$$a_{ij} = \frac{x_{ij}}{x_i}$$

In the next section, we will learn about a sampling method, MCMC, which is used for high-dimensional probability distributions wherein the next sample is dependent on the current sample drawn randomly from a population. In short, the samples drawn from the distribution are probabilistically dependent on each other. The volume of a sample space increases exponentially with the number of parameters or dimensions, and modeling such a space could easily be inaccurate with the usage of straightforward methods such as Monte Carlo sampling. The MCMC method is an attempt to harness the properties of a random problem and construct the corresponding Markov process efficiently.

Markov Chain Monte Carlo

MCMC is a method of random sampling from a target population/distribution defined by high-dimensional probability definition. It is a large-scale statistical method that draws samples randomly from a complex probabilistic space to approximate the distribution of attributes over a range of future states. It helps gauge the distribution of a future outcome and the sample averages help approximate expectations. A Markov chain is a **graph** of states over which a sampling algorithm takes a random walk.

The most known MCMC algorithm is perhaps Gibbs sampling. The algorithms are nothing but different methodologies for constructing the Markov chain. The most general MCMC algorithm is Metropolis-Hastings and has flexibility in many ways. These two algorithms will be discussed in the next subsections.

Gibbs sampling algorithm

In Gibbs sampling, the probability of the next sample in the Markov chain is calculated as the conditional probability of the prior sample. Samples in the Markov chain are constructed by changing one random variable at a time (conditioned on other variables in the distribution), meaning subsequent samples in the search space are closer. Gibbs sampling is most appropriate with a discrete (not continuous) distribution, which has a parametric form that allows sampling and calculating the conditional probability. An example of sampling with Gibbs sampler is shown in *Figure 8.4*, which reproduces the desired distribution.

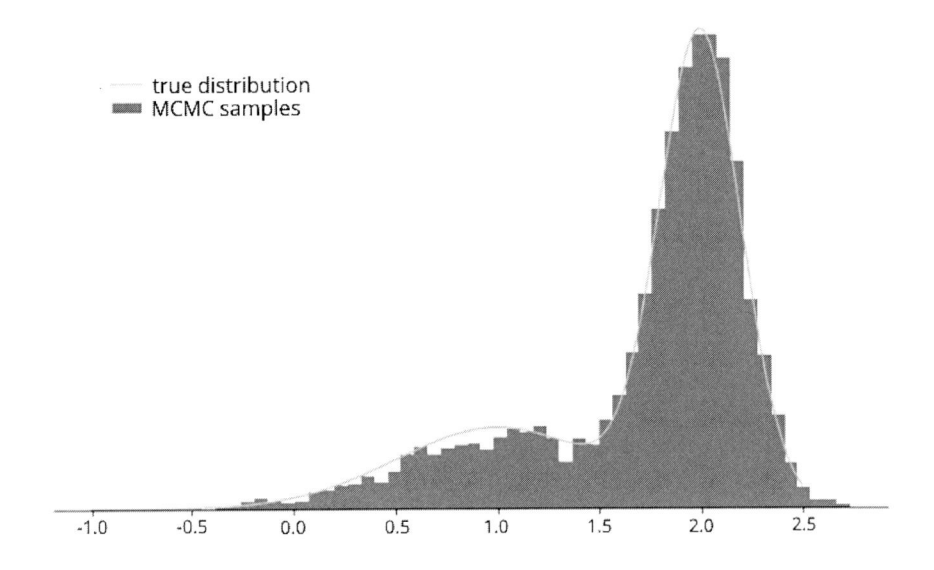

Figure 8.4: Gibbs sampler reproducing a desired Gaussian mixture

A Gibbs sampler is more efficient than a Metropolis-Hastings algorithm (discussed in the next subsection). It starts with a proposal distribution and a proposal is always accepted; that is, the acceptance probability is always 1. We will use an example of the bivariate Gaussian distribution to illustrate a Gibbs sampler with Python code in the last subsection.

Metropolis-Hastings algorithm

The Metropolis-Hastings algorithm is used for probabilistic models where Gibbs sampling cannot be used. It does not assume that the state of the next sample can be generated from a target distribution, which is the main assumption in Gibbs sampling. This algorithm involves using a surrogate probability distribution, also called the kernel, and an acceptance criterion that helps decide whether the new sample can be accepted into the Markov chain or has to be rejected. The proposed distribution (surrogate) is suggestive of an arbitrary next sample and the acceptance criterion ensures an appropriate limiting direction in getting closer to the true or desired the state of the next sample. The starting points of these algorithms are important and different proposal distributions can be explored.

How does this algorithm work?

1. We start with a random state.
2. Based on the proposal probability, we randomly pick a new state.
3. We calculate the acceptance probability of the proposed new state.

    For example, say the probability of a flipped coin landing on heads is the acceptance probability. If it lands on heads, we accept the sample; otherwise, we reject it.

4. We repeat the process for a long time.

We discard the initial few samples as the chain does not reach its stationary state. The period before the chain reaches its stationary state is called the burn-in period (see *Figure 8.5*). The accepted draws will converge to the stationary distribution after some time.

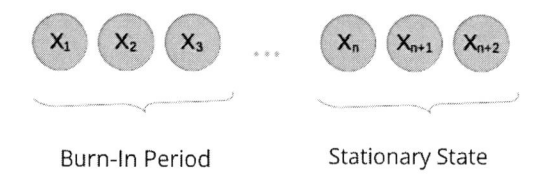

Figure 8.5: Markov chain

The stationary distribution shows the probability of being at any state X at any given time and is always reached if a very large number of samples is generated. This distribution is exactly the posterior distribution we're looking for. A posterior distribution is proportional to the product of likelihood and prior distribution. The Metropolis-Hastings algorithm is analogous to a diffusion process wherein all states are communicating (by design) and hence the system eventually settles into an equilibrium state, which is the same as converging to a stationary state. This property is called **ergodicity**.

In the next subsection, we illustrate the Metropolis-Hastings sampling algorithm, also with Python code, using the example of bivariate distribution.

Illustration of MCMC algorithms

The working of the Gibbs sampling algorithm is shown with a simple bivariate Gaussian distribution in the following code. We pass the two parameters (mu and sigma) for the conditional probability distribution and discard a part of initially sampled values for the algorithm to converge even if the starting (guess) value is way off. This part of the sample is known as burn-in:

```
import numpy as np
import matplotlib.pyplot as plt
import seaborn as sns
np.random.seed(42)
def gibbs_sampler(mus, sigmas, n_iter = 10000):
    samples = []
    y = mus[1]
    for _ in range(n_iter):
        x = p_x_y(y, mus, sigmas)
        y = p_y_x(x, mus, sigmas)
        samples.append([x, y])
    return samples
def p_x_y(y, mus, sigmas):
    mu = mus[0] + sigmas[1, 0]/sigmas[0, 0] * (y - mus[1])
    sigma = sigmas[0, 0]-sigmas[1, 0]/sigmas[1, 1]*sigmas[1, 0]
    return np.random.normal(mu, sigma)
def p_y_x(x, mus, sigmas):
    mu = mus[1] + sigmas[0, 1] / sigmas[1, 1]*(x - mus[0])
    sigma = sigmas[1, 1] - sigmas[0, 1]/sigmas[0, 0]*sigmas[0, 1]
    return np.random.normal(mu, sigma)
mus = np.asarray([5, 5])
sigmas = np.asarray([[1, 0.9], [0.9, 1]])
samples = gibbs_sampler(mus, sigmas)
burnin = 200
x = list(zip(*samples[burnin:]))[0]
y = list(zip(*samples[burnin:]))[1]
sns.jointplot(samples[burnin:], x = x, y = y, kind = 'kde')
sns.jointplot(samples[burnin:], x = x, y = y, kind = 'reg')
plt.show()
```

We run the code and the Gibbs sampler yields an output, shown in *Figure 8.6a*, in two forms, namely, a kernel distribution estimation plot and a linear regression fit. The output is the resulting (target) distribution based on sampled values using the Gibbs sampling algorithm.

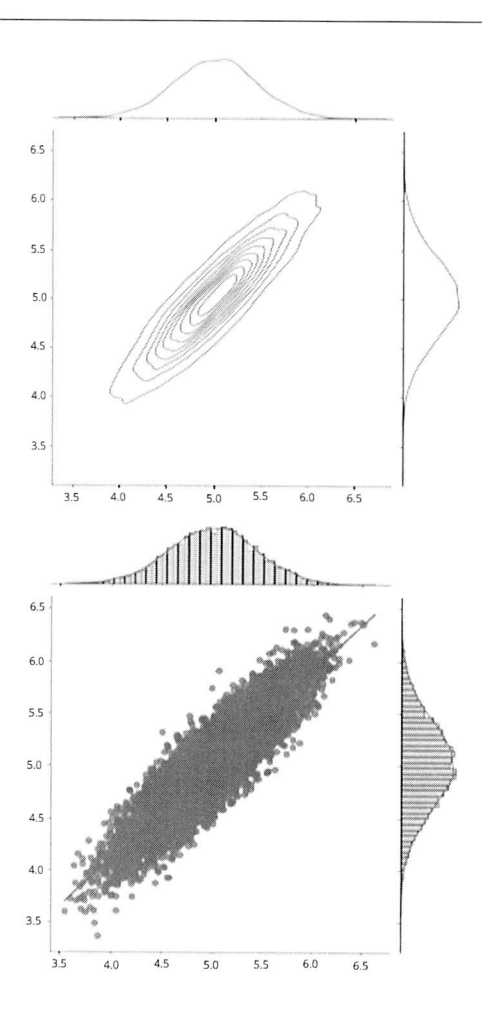

Figure 8.6a: Target distribution from the Gibbs sampling algorithm

We run a similar setup (bivariate distribution) for Metropolis-Hastings sampler. The Python code and output are given as follows. To begin with, we plot the true distribution and then use the multivariate normal distribution as the proposal. *Figure 8.6b* is the output (target distribution) based on sampling using the algorithm:

```
import numpy as np
import matplotlib.pyplot as plt
from tqdm import tqdm as tqdm
def density(z):
    z = np.reshape(z, [z.shape[0], 2])
    z1, z2 = z[:, 0], z[:, 1]
    norm = np.sqrt(z1 ** 2 + z2 ** 2)
```

```python
        exp1 = np.exp(-0.5 * ((z1 - 2) / 0.6) ** 2)
        exp2 = np.exp(-0.5 * ((z1 + 2) / 0.6) ** 2)
        v = 0.5 * ((norm - 2) / 0.4) ** 2 - np.log(exp1 + exp2)
        return np.exp(-v)
r = np.linspace(-5, 5, 1000)
z = np.array(np.meshgrid(r, r)).transpose(1, 2, 0)
z = np.reshape(z, [z.shape[0] * z.shape[1], -1])
def metropolis_sampler(target_density, size = 100000):
    burnin = 5000
    size += burnin
    x0 = np.array([[0, 0]])
    xt = x0
    samples = []
    for i in tqdm(range(size)):
        xt_candidate = np.array([np.random.multivariate_normal(xt[0], np.eye(2))])
        accept_prob = (target_density(xt_candidate))/(target_density(xt))
        if np.random.uniform(0, 1) < accept_prob:
            xt = xt_candidate
        samples.append(xt)
    samples = np.array(samples[burnin:])
    samples = np.reshape(samples, [samples.shape[0], 2])
    return samples
q = density(z) #true density
plt.hexbin(z[:,0], z[:,1], C = q.squeeze())
plt.gca().set_aspect('equal', adjustable ='box')
plt.xlim([-3, 3])
plt.ylim([-3, 3])
plt.show()
samples = metropolis_sampler(density)
plt.hexbin(samples[:,0], samples[:,1])
plt.gca().set_aspect('equal', adjustable = 'box')
plt.xlim([-3, 3])
plt.ylim([-3, 3])
plt.show()
```

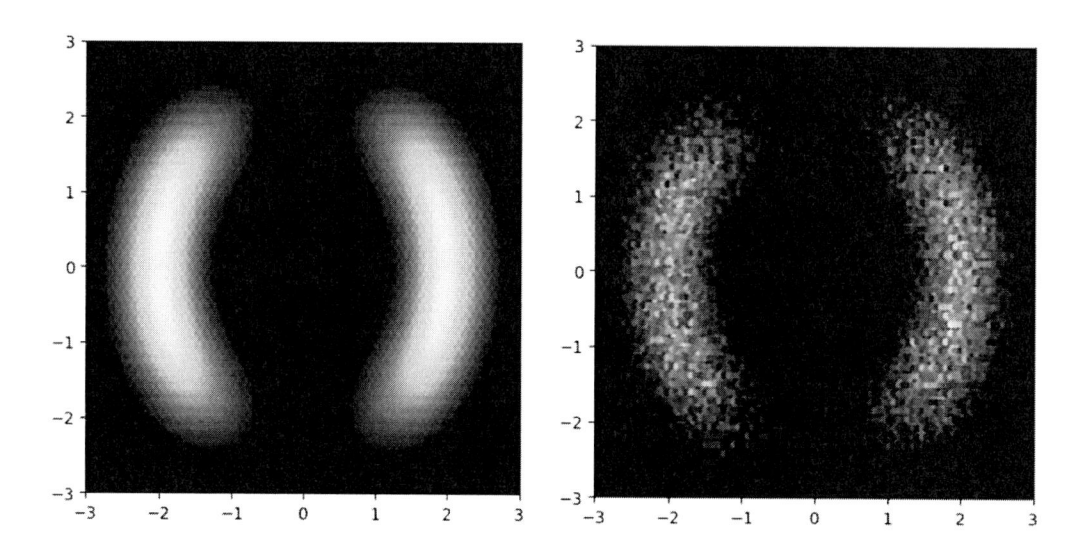

Figure 8.6b: True distribution (L) and target distribution (R) from
the Metropolis-Hastings sampling algorithm

For finite (discrete as well as continuous) state spaces, the existence of a unique stationary state is guaranteed. We start from a prior probability distribution and end with a stationary distribution, which is the posterior or target distribution based on sampled values.

Summary

In this chapter, we learned about the Markov chain, which is utilized to model special types of stochastic processes, such as problems wherein one can assume the entire past is encoded in the present, which in turn can be leveraged to determine the next (future) state. An application of the Markov chain in modeling time-series data was illustrated. The most common MCMC algorithm (Metropolis-Hastings) for sampling was also covered with code to illustrate. If a system exhibits non-stationary behavior (transition probability changes with time), then a Markov chain is not the appropriate model and a more complex model may be required to capture the behavior of the dynamic system.

With this chapter, we conclude the second part of the book. In the next chapter, we will explore fundamental optimization techniques, some of which are used in machine learning. We will touch upon evolutionary optimization, optimization in operations research, and that are leveraged in training neural networks.

Part 3: Mathematical Optimization

In this part, you will have exposure to optimization techniques that lay the foundation for machine learning, deep learning, and other models used in operations research. Optimization techniques are extremely powerful for predictive and prescriptive analytics and find applications in several complex problems in heavy industry. Additionally, blending classical mathematical modeling with machine learning often allows for the extraction of more meaningful insights for specific sensitive business problems.

This part has the following chapters:

- *Chapter 9, Exploring Optimization Techniques*
- *Chapter 10, Optimization Techniques for Machine Learning*

9
Exploring Optimization Techniques

This chapter primarily aims to address the question, "Why is optimization necessary while solving problems?" Mathematical optimization, or mathematical programming, is a powerful decision-making tool that has been talked about in depth in the chapters of Part I. What is important is to recall the simple fact that optimization yields the best result to a problem by reducing errors that are, essentially, the gaps between predicted and real data. Optimization comes at a cost; almost all optimization problems are described in terms of costs such as money, time, and resources. This cost function is the error function. If a business problem has clear goals and constraints, such as in the airline and logistics industries, mathematical optimization is applied for efficient decision-making.

In **machine learning** (**ML**) problems, the cost is often called the **loss function**. ML models make predictions about trends or classify data wherein training a model is a process of optimization, as each iteration in this process aims to improve the accuracy of the model and lower the margin of error. Selecting the optimum value of hyperparameters is key to ensuring an accurately and efficiently performing model. Hyperparameters are the elements of an ML model (for example, learning rate, number of clusters, etc.) that are tuned to fit a specific dataset to the model. In short, they are parameters whose values control the learning process. Optimization is an iterative process, meaning that the ML model becomes more accurate with each iteration in most cases and becomes better at predicting an outcome or classifying data.

The right blend of ML and mathematical optimization can prove to be useful for certain business problems. For example, the output of an ML model can determine the scope of an optimization model, especially in routing problems where one uses both predictive maintenance with ML as well as clustering, the results of which are fed into a mathematical model to create optimal routes. Similarly, an ML model may learn from a mathematical model. Initial values of decision variables obtained from a mathematical model can be used in an ML model that not only predicts optimal values of the decision variables, but also helps accelerate the performance of an optimization algorithm.

ML optimization is performed using algorithms that exploit a range of techniques to refine an ML model. The optimization process searches for the most effective configuration or set of hyperparameters for the model to suit the specific use case (dataset) or business problem.

To summarize, ML is data-driven and optimization is algorithm-driven. Every ML model operates on the principle of minimizing the cost function; hence, optimization is a superset at its core.

This chapter covers the following topics:

- Optimizing machine learning models
- Optimization in operations research
- Evolutionary optimization

The next section explores approaches and techniques used in optimizing ML models to arrive at the best set of hyperparameters.

Optimizing machine learning models

The concept of optimization is integral to an ML model. ML helps make clusters, detect anomalies, predict the future from historical data, and so forth. However, when it comes to minimizing costs in a business, finding optimal placement of business facilities, et cetera, what we need is a mathematical optimization model.

We will talk about optimization in machine learning in this section. Optimization ensures that the structure and configuration of the ML model are as effective as possible to achieve the goal it has been built for. Optimization techniques automate the testing of different model configurations. The best configuration (set of hyperparameters) has the lowest margin of error, thereby yielding the most accurate model for a given dataset. Getting the hyperparameter optimization right for an ML model can be tedious, as both under-optimized (underfit) as well as over-optimized (overfit) models fail. Overfitting is when a model is trained too closely to training data, resulting in inaccurate yields with new data. Underfitting is when a model is poorly trained, making it ineffective with training data as well as new data. Hyperparameters can be sought manually, which is an exhaustive method using trial and error. Underfit, optimal, and overfit models are illustrated in *Figure 9.1* as follows:

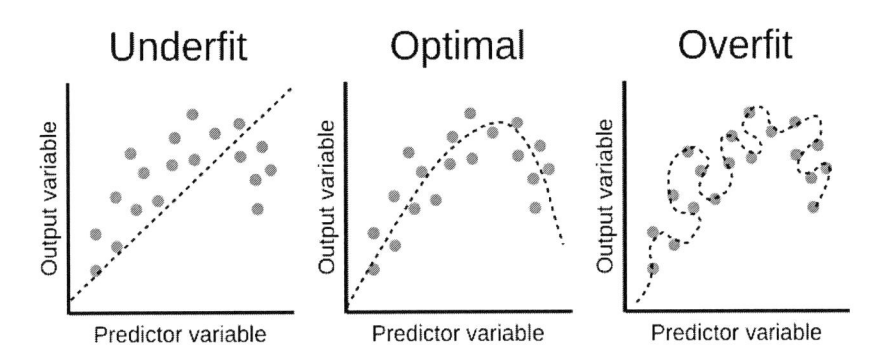

Figure 9.1: Under-optimized (L) and over-optimized (R) model fits

The main techniques of optimization include **random search**, **grid search** of hyperparameters, and **Bayesian optimization**, all of which are discussed in the following subsections.

Random search

The process of random sampling of the search space and identifying the most effective configuration of a hyperparameter set is random search. A random search technique discovers new combinations of hyperparameters for an optimized ML model. The number of iterations in the search process has to be set, which limits these new combinations, without which the process is a lot more time-consuming. It is an efficient process as it replaces an exhaustive search with randomness. A search space can be thought of as a volume in space, each dimension of which represents a hyperparameter, and each point or vector in the volume represents a model configuration. An optimization procedure involves defining the search space.

The search space is a dictionary in the Python code, and the `scikit-learn` library provides functions to tune model hyperparameters. An example code of a random search for a classification model is provided here:

```
import pandas as pd
from scipy.stats import loguniform
from sklearn.linear_model import LogisticRegression
from sklearn.model_selection import RepeatedStratifiedKFold
from sklearn.model_selection import RandomizedSearchCV

dataframe = pd.read_csv('sonar.csv')
data = dataframe.values
X, y = data[:, :-1], data[:, -1]
#Model
model = LogisticRegression()
cv = RepeatedStratifiedKFold(n_splits = 10, n_repeats = 3, random_
```

```
state = 1)
#Define search space
space = dict()
space['solver'] = ['newton-cg', 'lbfgs', 'liblinear']
space['penalty'] = ['none', 'l1', 'l2', 'elasticnet']
space['C'] = loguniform(1e-5, 100)

search = RandomizedSearchCV(model, space, n_iter = 500, scoring = 
'accuracy',
                            n_jobs = -1, cv = cv, random_state = 1)
result = search.fit(X, y)

print('Best Score: %s' % result.best_score_)
print('Best Hyperparameters: %s' % result.best_params_)
```

The dataset used is a set of 60 patterns obtained by bouncing sonar signals off a metal cylinder under various conditions. Each pattern is a set of numbers lying in the range between 0.0 and 1.0, with each number representing the energy within a frequency band integrated over a period of time. The label associated with each record is either *R* if the object is a rock, or *M* if the object is a metal cylinder or mine. Data can be found in the GitHub repository at https://github.com/ranja-sarkar/dataset.

An example code of random search for a linear regression model has also been provided. The insurance dataset with two variables, namely the number of claims and total payment (in Swedish Krona) for all claims in geographical Swedish zones, can be found in the GitHub repository at https://github.com/ranja-sarkar/dataset.

The difference between regression and classification tasks is in choosing the performance scoring protocol for the models. The hyperparameter optimization methods in the `scikit-learn` Python library assume good performance scores are negative values close to zero (for regression), with zero representing a perfect regression model:

```
import pandas as pd
from scipy.stats import loguniform

from sklearn.linear_model import Ridge
from sklearn.model_selection import RepeatedKFold
from sklearn.model_selection import RandomizedSearchCV

df = pd.read_csv('auto-insurance.csv')
data = df.values
X, y = data[:, :-1], data[:, -1]
#Model
model = Ridge()
```

```
cv = RepeatedKFold(n_splits = 10, n_repeats = 3, random_state = 1)

#Define search space
space = dict()
space['solver'] = ['svd', 'cholesky', 'lsqr', 'sag']
space['alpha'] = loguniform(1e-5, 100)
space['fit_intercept'] = [True, False]
space['normalize'] = [True, False]

search = RandomizedSearchCV(model, space, n_iter = 500, scoring
=       'neg_mean_absolute_error', n_jobs = -1, cv = cv, random_state =
1)
result = search.fit(X, y)
print('Best Score: %s' % result.best_score_)
print('Best Hyperparameters: %s' % result.best_params_)
```

The runtime of the code depends on the size of the search space and the system processor speed. The `result` class in the code provides the outcome, the most important value being the best score for the best performance of the model and the hyperparameters that achieved this score. Once the best set of hyperparameters becomes known, one can define a new model, set the hyperparameters to the known values, and fit the model on available data. This model can then be used for predictions on new data. The number of random configurations in the parameter space look like *Figure 9.2*, which shows that random search works best for low-dimensional data:

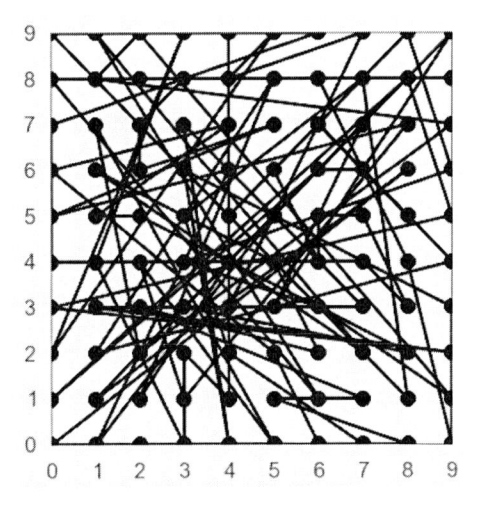

Figure 9.2: Random search

In the next subsection, we elaborate on grid search for optimization of classification and regression models.

Grid search

The process of assessing the effectiveness of known hyperparameter values of an ML model is grid search. Each hyperparameter is represented as a dimension on a grid across the search space and each point in the grid is searched and evaluated. Grid search is great for checking intuitive guesses and hyperparameter combinations that are known to perform well in general. As mentioned earlier, an optimization procedure involves defining a search space (a dictionary in Python), which can be thought of as a volume where each dimension represents a hyperparameter and each point (vector) represents a model configuration. A discrete grid has to be defined here. In other words, the grid search space takes discrete values (that can be on a log scale) instead of a log-uniform distribution used in a random search space.

A sample code of grid search for a classification model using the same dataset (sonar.csv) explored for a random search algorithm is given here:

```
import pandas as pd
from scipy.stats import loguniform
from sklearn.linear_model import LogisticRegression
from sklearn.model_selection import RepeatedStratifiedKFold
from sklearn.model_selection import RandomizedSearchCV


dataframe = pd.read_csv('sonar.csv')
data = dataframe.values
X, y = data[:, :-1], data[:, -1]

#Model
model = LogisticRegression()
cv = RepeatedStratifiedKFold(n_splits = 10, n_repeats = 3, random_state = 1)
#Define search space
space = dict()
space['solver'] = ['newton-cg', 'lbfgs', 'liblinear']
space['penalty'] = ['none', 'l1', 'l2', 'elasticnet']
space['C'] = [1e-5, 1e-4, 1e-3, 1e-2, 1e-1, 1, 10, 100]

search = GridSearchCV(model, space, scoring = 'accuracy', n_jobs = -1, cv = cv)
result = search.fit(X, y)

print('Best Score: %s' % result.best_score_)
print('Best Hyperparameters: %s' % result.best_params_)
```

A sample code of a grid search for a linear regression model using the same dataset (auto-insurance.csv) explored for a random search algorithm is provided as follows. The best hyperparameters obtained using the random search and grid search algorithms for this dataset can be compared to get an estimate of which algorithm works better for the dataset:

```
import pandas as pd
from sklearn.linear_model import Ridge
from sklearn.model_selection import RepeatedKFold
from sklearn.model_selection import GridSearchCV
df = pd.read_csv('auto-insurance.csv')
data = df.values
X, y = data[:, :-1], data[:, -1]
#Model
model = Ridge()
cv = RepeatedKFold(n_splits = 10, n_repeats = 3, random_state = 1)
#Define search space
space = dict()
space['solver'] = ['svd', 'cholesky', 'lsqr', 'sag']
space['alpha'] = [1e-5, 1e-4, 1e-3, 1e-2, 1e-1, 1, 10, 100]
space['fit_intercept'] = [True, False]
space['normalize'] = [True, False]

search = GridSearchCV(model, space, scoring = 'neg_mean_absolute_error', n_jobs = -1, cv = cv)
result = search.fit(X, y)
print('Best Score: %s' % result.best_score_)
print('Best Hyperparameters: %s' % result.best_params_)
```

The scores obtained for the datasets with random search and grid search in classification and regression models are nearly identical. The selection of optimization technique for a given dataset depends on the use case. Though random search might in some cases result in better performance, it needs more time, while grid search is appropriate for quick searches of hyperparameters that perform well in general. The values of hyperparameters are placed like a matrix as shown in *Figure 9.3*, similar to a grid:

102   Exploring Optimization Techniques

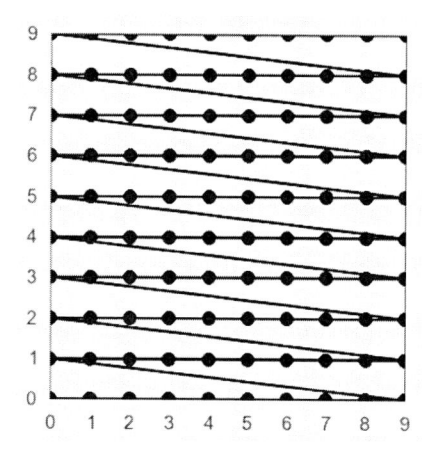

Figure 9.3: Grid search

Another method, known as Bayesian optimization, whose search procedure is different from the preceding two, is discussed in the following subsection.

Bayesian optimization

A directed and iterative approach to global optimization using probability is Bayesian optimization. This is a Gaussian process that converges fast for continuous hyperparameters that is, in a continuous search space (*Figure 9.4*). In Bayesian optimization, a probabilistic model of the function is built, and maps hyperparameters to the objectives evaluated on a validation dataset. This process evaluates a hyperparameter configuration based on the current model, then updates it until an optimal point is reached and it attempts to find the global optimum in a minimum number of steps. In most cases, it is more efficient and effective than optimization by way of random search. The optimization landscape (multiple local minima) with one global minimum is illustrated as follows:

Figure 9.4: Optimization landscape (response surface)

Bayesian optimization incorporates prior belief (marginal probability) about the objective function and updates the prior with samples drawn from the function to obtain a posterior belief (conditional probability) that better approximates the function, which is illustrated in *Figure 9.5*. This process repeats itself until the extremum of the objective function is located or resources are exhausted:

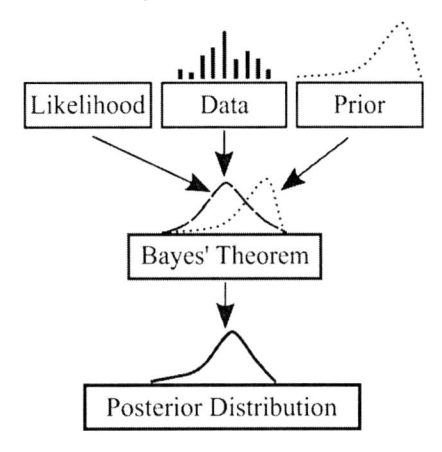

Figure 9.5: Bayesian statistics

Bayesian search is typically beneficial when there is a large amount of data, the learning is slow, and tuning time has to be minimized. The `scikit-optimize` library provides functions for Bayesian optimization of ML models. A sample code for hyperparameter tuning by the Bayesian method in a classification problem is provided as follows:

```
import numpy as np
from sklearn.datasets import make_blobs
from sklearn.model_selection import cross_val_score
from sklearn.neighbors import KNeighborsClassifier
from skopt.space import Integer
from skopt.utils import use_named_args
from skopt import gp_minimize

#Generate classification dataset
X, y = make_blobs(n_samples = 500, centers = 3, n_features = 2) ##3 class labels in data
#Model kNN
model = KNeighborsClassifier()
#Define search space
search_space = [Integer(1, 5, name = 'n_neighbors'), Integer(1, 2, name = 'p')]
@use_named_args(search_space)
def evaluate_model(**params):
```

```
        model.set_params(**params)
        result = cross_val_score(model, X, y, cv = 5, n_jobs = -1, scoring
= 'accuracy')
        estimate = np.mean(result)
        return 1.0 - estimate
#Optimize
result = gp_minimize(evaluate_model, search_space)
print('Accuracy: %.3f' % (1.0 - result.fun))
print('Best Parameters: n_neighbors = %d, p = %d' % (result.x[0],
result.x[1]))
```

The model used for approximating the objective function is called the **surrogate model**, and the posterior probability is a surrogate objective function that can be used to estimate the cost of candidate samples. The posterior is used to select the next sample from the search space and the technique that does this is called the **acquisition function**. Bayesian optimization is best when the function evaluation is expensive or the form of the objective function is complex (nonlinear, non-convex, highly multi-dimensional, or highly noisy) – for example, in deep neural networks.

The process of optimization lowers errors or loss from predictions in an ML model and improves the model's accuracy. The very premise of ML relies on a form of function optimization so inputs can be almost accurately mapped to expected outputs.

In the next section, we will learn about mathematical optimization in operations research.

Optimization in operations research

The term **operations research** was coined during World War I, when the British military brought together a group of scientists to allocate insufficient resources (food, medicines, etc.) in the most effective way possible to different military operations. Therefore, the term implies optimization, which is maximizing or minimizing an objective function subject to constraints, often in complex problems and in high dimensions. Operations problems typically include planning work shifts or creating a schedule for large organizations, designing facilities for customers at a large store, choosing investments for available funds, supply chain management, and inventory management, all of which can be posed or formulated as mathematical problems with a collection of variables and their relationships.

In operations research, a business problem is mapped to a lower-level generic problem that is concise enough to be described in mathematical notations. These generic problems can in turn be expressed using higher-level languages; for example, resources and activities are used to describe a scheduling problem. The higher-level language is problem-specific, hence the generic problems can be described using modeling paradigms. A modeling paradigm is a set of rules and practices that allows for the representation of higher-level problems using lower-level data structures such as matrices. These data structures or matrices are passed to the last step of abstraction, which is algorithms. The most prominent modeling paradigms are linear programming, integer programming, and mixed-integer programming, all of which use linear equality constraints. There is a family of algorithms to solve these linear programming problems. Search algorithms, such as branch and bound, solve integer programming problems, while the simplex algorithm is used in a linear programming modeling paradigm.

An example of how to solve a knapsack problem by optimization is illustrated with the following data (*Figures 9.6a and 9.6b*):

id	Item	Weight in kg	Value
1	Sleeping bag	1.2	5
2	Pillow	0.39	2
3	Torch	0.5	5
4	First Aid Kit	0.5	4
5	Hand sanitiser	0.5	1

Figure 9.6a: Knapsack problem

Let's say the constraint is the ability to only carry a maximum of 2.9 kg in the camping sack, while the total weight of all items is 3.09 kg. The item's value assists in choosing the optimum number of items. As the number of items increases, the problem becomes bigger, and solving it by trying all possible combinations of items takes a significant amount of time:

id	Item	Weight in kg	Value	Value per weight
1	Sleeping bag	1.2	5	4.17
2	Pillow	0.39	2	5.13
3	Torch	0.5	5	10.00
4	First Aid Kit	0.5	4	8.00
5	Hand sanitiser	0.5	1	2.00

Figure 9.6b: Knapsack problem with another variable

The objective function is value, which must be maximized. The best of items has to be chosen to meet the constraint of 2.9 kg by total weight. A solver (pulp, in this case) is used to solve this linear programming problem, as shown in the following code. The decision variables (to be determined) are given by $x_i = \{1, 0\}$. The variable is 1 if the item is chosen and 0 if the item is not chosen:

```
from pulp import *
#value per weight
v = {'Sleeping bag': 4.17, 'Pillow': 5.13, 'Torch': 10.0, 'First Aid
Kit': 8.0, 'Hand sanitiser': 2.0}
#weight
w = {'Sleeping bag': 1.2, 'Pillow': 0.39, 'Torch': 0.5, 'First Aid
Kit': 0.5, 'Hand sanitiser': 0.5}

limit = 2.9
items = list(sorted(v.keys()))
```

```
# Model
m = LpProblem("Knapsack Problem", LpMaximize)
# Variables
x = LpVariable.dicts('x', items, lowBound = 0, upBound = 1, cat = 
LpInteger)

#Objective
m += sum(v[i]*x[i] for i in items)
#Constraint
m += sum(w[i]*x[i] for i in items) <= limit
#Optimize
m.solve()
#decision variables
for i in items:
    print("%s = %f" % (x[i].name, x[i].varValue))
```

This code when executed results in the following output:

```
x_First_Aid_Kit = 1.0
x_Hand_sanitizer = 0.0
x_Pillow = 1.0
x_Sleeping_bag = 1.0
x_Torch = 1.0
```

Going by the result (optimal solution), a hand sanitizer must not be carried in the sack. This is a simple integer programming problem as decision variables are restricted to being integers. In a very similar manner, other practical business problems such as production planning are solved by mathematical optimization wherein the right resources are chosen to maximize profit and so on. When operations research is combined with ML predictions, data science is effectively transformed into decision science, allowing organizations to make actionable decisions.

In the next section, we will learn about **evolutionary optimization**, which is motivated by optimization processes observed in nature such as the migration of species, bird swarms, and ant colonies.

Evolutionary optimization

Evolutionary optimization makes use of algorithms that mimic the selection process within the natural world. Examples of this are genetic algorithms that optimize via natural selection. Each iteration of a hyperparameter value is like a mutation in genetics that is assessed and altered. The process continues using recombined choices until the most effective configuration is reached. Hence, each generation improves with every iteration as it is optimized. Genetic algorithms are often used to train neural networks.

An evolutionary algorithm typically consists of three steps: initialization, selection, and termination. Fitter generations survive and proliferate, like in natural selection. In general, an initial population of a wide range of solutions is randomly created within the constraints of the problem. The population contains an arbitrary number of possible solutions to the problem, or the solutions are roughly centered around what is believed to be an ideal solution. These solutions are then evaluated in the next step according to a fitness (or objective) function. A good fitness function is one that represents the data and calculates a numerical value for the viability of a solution to a specific problem. Once the fitness of all solutions is calculated, the top-scoring solutions are selected. There may be multiple fitness functions that result in more than one optimal solution, which is when a decider is used to narrow down a single problem-specific solution that is based on some key metrics. *Figure 9.7* depicts the steps of these algorithms as follows:

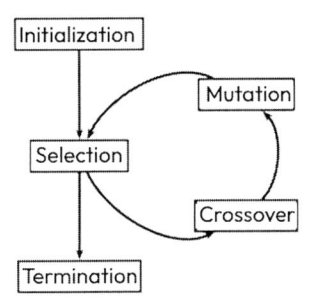

Figure 9.7: Steps of evolutionary algorithms

The top solutions make the next generation in the algorithm. These solutions typically have a mixture of the characteristics of solutions from the previous generation. New genetic material is introduced into this new generation, which, mathematically speaking, means introducing new probability distribution. This step is mutation, without which optimal results are difficult to achieve. The last step is termination, when the algorithm reaches either some threshold of performance or some maximum number of iterations (runtime). A final solution is then selected and returned.

An evolutionary algorithm is a heuristic-based approach to solving problems that would take too long to exhaustively process using deterministic methods. It is a stochastic search technique typically applied to combinatorial problems or in tandem with other methods to find an optimal solution quickly.

Summary

In this chapter, we learned about optimization techniques, especially the ones used in machine learning that aim to find the most effective hyperparameter configuration for an ML model fitted to a dataset. An optimized ML model has minimum errors, thereby improving the accuracy of predictions. There would be no learning or development of models without optimization.

We touched upon optimization algorithms that are used in operations research, as well as evolutionary algorithms that find usage in the optimization of deep learning models and network modeling of more complex problems.

In the final chapter of the book, we will learn about how standard techniques are selected to optimize ML models. Multiple optimal solutions may exist for a given problem and there may be multiple optimization techniques to arrive at them. Hence, it is essential to choose the technique carefully while building the model addressing the pertinent business question.

10
Optimization Techniques for Machine Learning

We discussed mathematical optimization techniques in the previous chapter and their necessity in business problems that require minimizing the cost (error) function and in predictive modeling, wherein the machine learns from historical data to predict the future. In **Machine Learning** (**ML**), the cost is a loss function or an energy function that is minimized. It can be challenging in most cases to know which optimization algorithm should be considered for a given ML model. Optimization is an iterative process to maximize or minimize an objective function and there is always a trade-off between the number of iteration steps taken and the computational hardship to get to the next step. In this chapter, hints of how to choose an optimization algorithm given a problem (hence, an objective) have been provided. The choice of optimization algorithm depends on different factors, including the specific problem to be solved, the size and complexity of the associated dataset, and the resources, such as computational power and memory, available.

Direct search as well as stochastic search algorithms are designed for an objective function where the derivative of this function is not available. Strictly speaking, optimization algorithms can be grouped into those that use derivatives and those that do not use derivatives. Optimization algorithms that rely on gradient descent are fast and efficient; however, they require well-behaved objective functions to work well. We can fall back on an exhaustive search if the function has tricky characteristics, but it takes an extremely long time (*Figure 10.1*). There are optimization methods tougher than gradient descent, such as **Genetic Algorithms** (**GAs**) and simulated annealing. These take longer computational time and a greater number of steps than gradient descent, but they discover the optimal point even when it is very difficult to find.

110   Optimization Techniques for Machine Learning

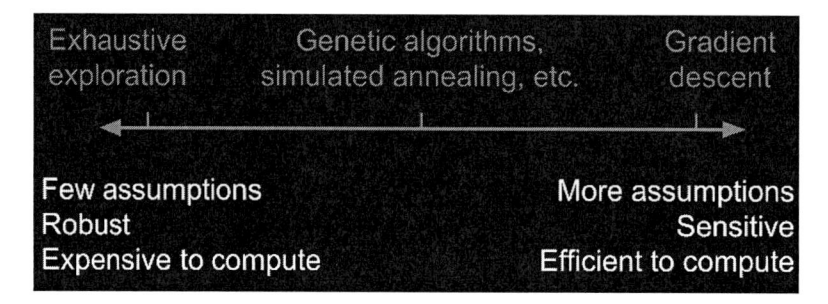

Figure 10.1: Performance of optimization algorithms

There can be derivative-free as well as gradient-based algorithms for optimization. Optimization algorithms used in ML models can in general be grouped into ones that use the first derivative (called the gradient) of the objective function and others that use the second derivative (called the Hessian) of the function in the search space.

This chapter covers the following topics:

- General optimization algorithms
- Complex optimization algorithms

Complex optimization algorithms encompass differentiable and non-differential functions. The next two sections cover examples of general and complex optimization algorithms.

General optimization algorithms

The most common optimization problem encountered in ML is continuous function optimization, wherein the function's input arguments are (real) numeric values. In training ML models, optimization entails minimizing the loss function till it reaches or converges to a local minimum (value).

An entire search domain is utilized in global optimization whereas only a neighborhood is explored in local optimization, which requires the knowledge of an initial approximation, as evident from *Figure 10.2a*. If the objective function has local minima, then local search algorithms (gradient methods, for example) can also be stuck in one of the local minima. If the algorithm attains a local minimum, it is nearly impossible to reach the global minimum in the search space. In discrete search space, the neighborhood is a finite set that can be completely explored, while the objective function is differentiable (gradient methods, Newton-like methods) in continuous search space.

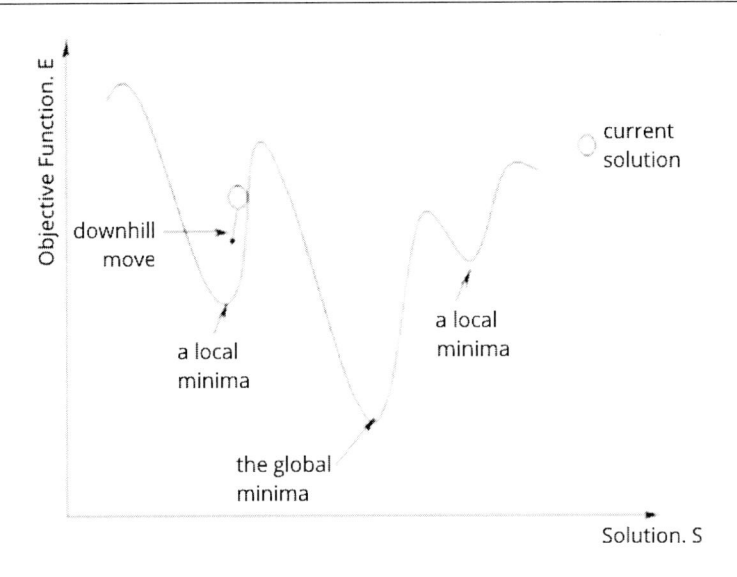

Figure 10.2a: Local minimum versus global minimum

Functions may be of a discrete nature, taking discrete variables, and are found in combinatorial optimization problems (an example is the **Traveling Salesman Problem** (**TSP**), depicted in *Figure 10.2b*) wherein the feasible solutions are also discrete. Generally speaking, it is more efficient searching through continuous space to find the optimum than searching through discrete space.

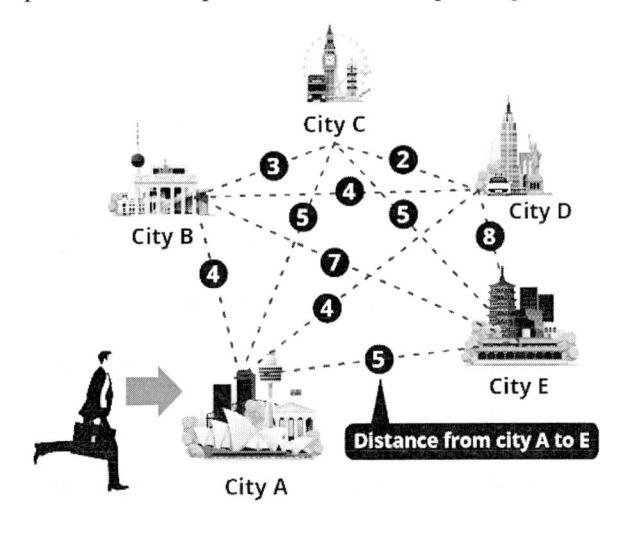

Figure 10.2b: TSP is a combinatorial optimization problem

Bracketing algorithms are optimization algorithms with one input variable where the optima is known to exist within a specific range. They assume that a single optimum (unimodal objective function) is present in the known range of search space. These algorithms may sometimes even be used when the derivative information is unavailable. The bisection method of optimization is one such example.

Optimization algorithms with more than one input variable are local descent algorithms. The process in local descent involves choosing a direction for movement in the search space, then performing a bracketing search in a line or hyperplane in the chosen direction. Local descent is also called the line search algorithm; it is, however, computationally expensive to optimize each directional move in the search space. Gradient descent is a classic example of the line search algorithm.

Algorithms that are grouped in accordance with whether they use gradient (first-order) or gradient of gradient (second-order) information to move in the search space to find the optimal point are discussed in the following subsections.

First-order algorithms

The first derivative (gradient or slope) of the objective function is used in first-order optimization algorithms. First-order algorithms are generally referred to as gradient descent (or steepest descent). Unlike gradient descent, regularization algorithms use a predefined objective function. An ML model learns by minimizing an objective (cost function) and regularization is used on top of that when such a model overfits.

The gradient in the search space is calculated using a step size, called the learning rate, which is a hyperparameter controlling the distance of movement in the space (*Figure 10.3*). Too small a step size leads to a long time to search for the optimum point, while too large a step size might lead to completely missing it. Optimizers have hyperparameters such as the learning rate, which can have a big impact on the performance of the ML model.

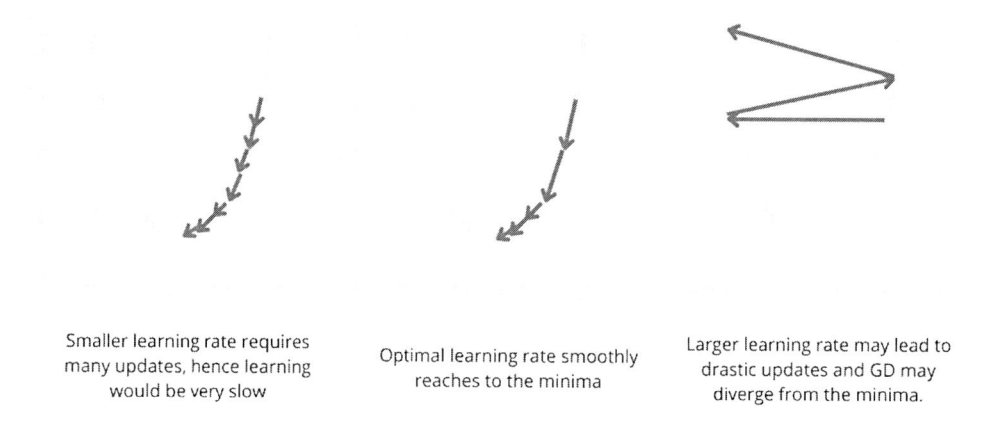

Figure 10.3: Learning rates in gradient descent

Gradient descent variants are batch gradient descent, mini-batch gradient descent, and **Stochastic Gradient Descent** (**SGD**). Batch gradient descent computes the gradient with respect to the entire training dataset (all training examples), whereas SGD computes that with respect to each training example. A mini-batch performs an update for every (mini-) subset of training examples and hence takes the best of both worlds. A batch gradient descent can be very slow, whereas mini-batch gradient descent is very efficient. Mini-batch gradient descent is a good choice for problems with huge data. SGD performs frequent updates and hence the objective function fluctuates heavily, but it brings better convergence to the optimum. SGD is used to train artificial neural networks.

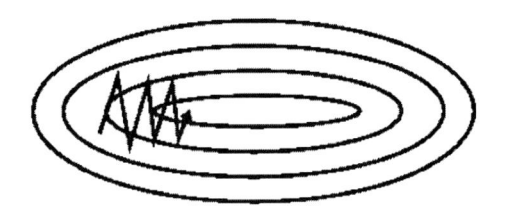

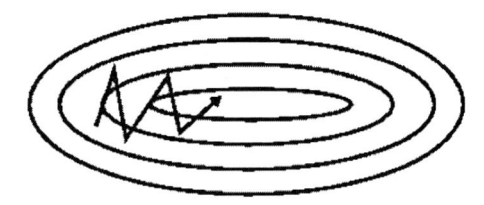

Image 2: SGD without momentum   Image 3: SGD with momentum

Figure 10.4: First-order algorithm (gradient descent) example

Minor extensions to the gradient descent procedure of optimization lead to several algorithms, such as momentum, **Adaptive Gradient** (**AdaGrad**), and **Adaptive Moment Estimation** (**Adam**). Momentum, for example, is a method that helps accelerate SGD in the relevant direction (*Figure 10.4*) for faster convergence. Methods such as adagrad and Adam compute adaptive learning rates for each parameter, helping the function converge quickly. However, Adam might be the best choice for sparse data. Adam uses both the gradient and second moment of the gradient. Adagrad is good for problems with very noisy data and ill-conditioned cost functions; that is; different dimensions of the cost function are not of the same scale.

Second-order algorithms

The second derivative (Hessian) of the objective function is used in second-order optimization algorithms, provided the Hessian (curvature) matrix can be either calculated or approximated. These algorithms are used for univariate objective functions that have a single real variable, few of which show either the minimum or the maximum while optimizing but a saddle point in its domain (search space). Newton's method is an example of a second-order optimization algorithm. A comparison of gradient descent (first-order) with Newton's method (second-order) of optimization is shown in *Figure 10.5*.

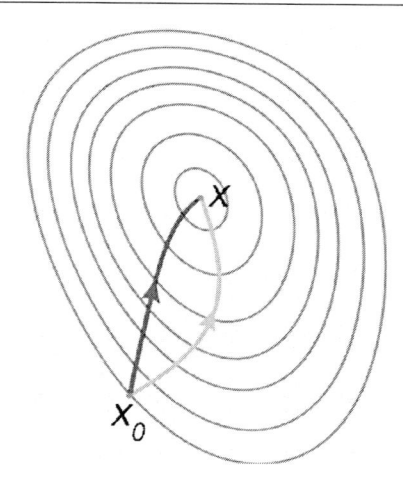

114   Optimization Techniques for Machine Learning

Figure 10.5: Gradient descent (green) and Newton's method (red) t,
to find routes from x_0 to x considering very small learning rates

Such algorithms work better for neural networks; however, computation and storage become challenging with a huge number of dimensions or parameters. In order to successfully use second-order algorithms, one must simplify the matrix, which is typically done by approximating the Hessian matrix with a simpler form.

The following section elaborates the differentiability of objective functions, which is what decides whether to select a general (discussed in this section) or complex optimization algorithm given a problem.

Complex optimization algorithms

The nature of the objective function helps select the algorithm to be considered for the optimization of a given business problem. The more information that is available about the function, the easier it is to optimize the function. Of most importance is the fact that the objective function can be differentiated at any point in the search space.

Differentiability of objective functions

A differentiable objective function is one for which the derivative can be calculated at any given point in input space. The derivative (slope) is the rate of change of the function at that point. The Hessian is the rate at which the derivative of the function changes. Calculus helps optimize simple differentiable functions analytically. For differentiable objective functions, gradient-based optimization algorithms are used. However, there are objective functions for which the derivative cannot be computed, typically for very complex (noisy, multimodal, etc.) functions, which are called non-differentiable objective functions. There can be discontinuous objective functions as well, for which the derivatives can only be calculated in some regions of the search space. Stochastic and population algorithms handle such functions and are, hence, sometimes called black-box algorithms.

When an analytical form of the objective function is not available, one generally uses simulation-based optimization methods. The next subsection talks briefly about the algorithms considered while finding a feasible solution is challenging using classical methods, and they either compute or build around assumptions about the derivatives of objective functions.

Direct and stochastic algorithms

Direct and stochastic optimization algorithms are used in problems where the derivative of the objective function is unknown or cannot be calculated, that is, the search space is discontinuous. The former algorithms are deterministic and assume the objective function is unimodal (it has a single global optimum). Direct search is often referred to as pattern search as it effectively navigates through the search space using geometric shapes. Gradient information is approximated from the objective function and used in initiating a line search in the search space, eventually (with repeated line searches) triangulating the region of optimum. Powell's method is one example of a direct search algorithm. It is a gradient-free method because the function to be optimized with it is non-differentiable.

On the other hand, stochastic algorithms make use of randomness in the global search, hence the name. These typically involve sampling the objective function and can handle problems with deceptive local optima. Simulated annealing (*Figure 10.6*) is an example of a stochastic search algorithm, that is, of global optimization, which occasionally accepts poorer initial configurations. Simulated annealing is a probabilistic technique used to solve unconstrained and bound-constrained optimization problems. It is a metaheuristic to approximate global optimization in a large search space of a physical process wherein the system energy is minimized.

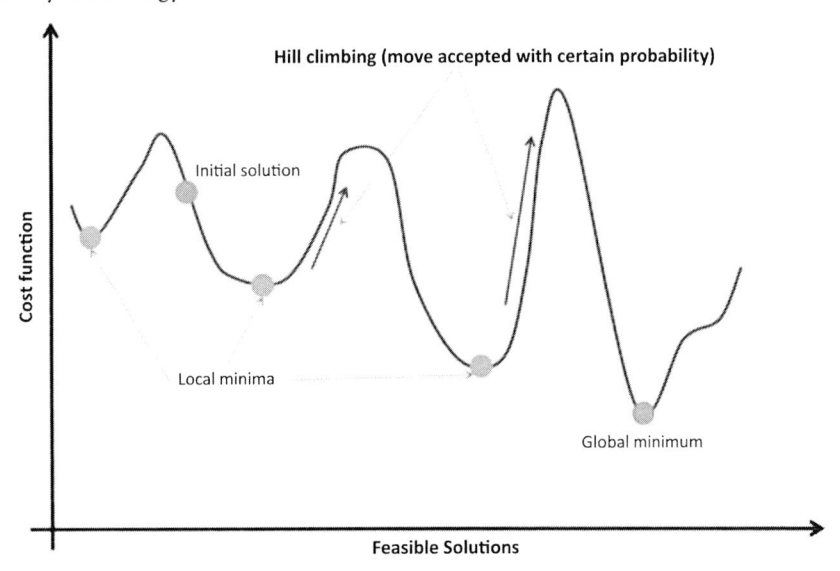

Figure 10.6: Simulated annealing is a stochastic optimization algorithm

Population optimization algorithms such as GAs are also stochastic and typically used for multimodal objective functions with multiple global optima and not-so-smooth (highly noisy) functions. These algorithms maintain a population of candidate solutions that add robustness to the search, thereby increasing the likelihood of overcoming local optima. The efficiency of these is very sensitive to the variables used in describing the problem. As with other heuristic algorithms, evolutionary algorithms have many degrees of freedom and, therefore, are difficult to tune for good model performance.

A GA pursues the evolution analogy, which proceeds by maintaining an even number of individuals in the population. These individuals make a generation, and a new generation is produced by randomly selecting a pair wherein the fitter individual is more likely to be chosen. GA is used to solve complex optimization problems by initialization (of the population), fitness assignment (to individuals in the population), and selection of the best (recombined) solution to the problem. A large community of researchers is working on GAs for utilization in most practical problems.

Summary

In this chapter, we gained knowledge about which optimization algorithm must be considered to minimize (continuous) objective functions that are generally encountered in ML models. Such models have a real-valued evaluation of the input variables and involve local search. The differentiability of an objective function is perhaps the most important factor when considering the optimization algorithm type for a given problem.

The chapter did not contain an exhaustive list of algorithms to optimize ML models but captured the essence of the main ones and their underlying behavior with examples. It also touched upon the concepts of deterministic optimization and stochastic optimization, the latter encompassing GAs, whose utility is evolving in real-world problems.

Epilogue

This book was primarily targeted at data scientists early in their careers. It was assumed that readers of this book have knowledge of linear algebra and the basics of statistics, differential equations, fundamental numerical algorithms, data types, and data structures. Having said that, it must be realized that transforming a business problem into a mathematical formulation is an art.

While exploring the world of data science, it is important to understand the relevance of classical mathematical models and how they can be utilized along with ML models to solve business problems, often complex ones. Hybrid (blended) models enable better decision-making and become particularly essential for high-stake decisions in sensitive domains. Mathematical optimization typically elevates an ML model for the best interpretation of the connection between decision variables and relevant data and business objectives and of the optimal solution to the business problem. Nevertheless, simpler (pure or unblended) models are more often explainable, and while building complex ones, we need to look at the aspects of efficiency and cost.

I would like to wind this book up by acknowledging and sincerely thanking the following subject matter experts:

- Brandon Rohrer (`https://github.com/brohrer`)
- Sebastian Raschka (`https://github.com/rasbt`)
- Jason Brownlee (`https://www.linkedin.com/in/jasonbrownlee/`)

Their online articles, books, courses, and tutorials/blogs have motivated me to learn, relearn, and deep dive into the world of data science and mathematical optimization. My learning and work experience eventually have taken shape in this book.

Index

A

acquisition function 104
adagrad method 47
Adaptive Moment Estimation (Adam) 18, 48, 113
anomaly detection , 39
artificial intelligence (AI) 13
automated machine learning (AutoML) 22

B

backpropagation 20
batch gradient descent (BGD) 45
Bayesian optimization 102, 103, 104
binary classifier 51
burn-in period 88

C

complex optimization algorithms 114
    differentiable objective function 114
    direct and stochastic algorithms 115, 116
control theory 10, 11
    problem 11, 12
    problem, formulation 13

covariance matrix 31
cross-entropy 17
curse of dimensionality 29

D

dataset
    determining, with number of PCs 32, 33
decision boundary 52
decision variables 4
deep learning (DL) 69
differentiable objective function 114
dimensionality reduction
    feature extraction 34
    feature selection 34
direct and stochastic algorithms 115, 116
directed acyclic graphs (DAGs) 62
directed graphs 64
Discrete Fourier Transform (DFT) 8
Discrete-Time Signal Processing (DSP) 8
dynamic model 73

E

eigendecomposition 31
eigenvalues 31
eigenvectors 31

ergodicity 88
estimate
    high accuracy and high precision 73
evolutionary optimization 106, 107
extreme gradient boost (XGB) 51

F

Fast Fourier Transform (FFT) 8
first-order algorithms 113

G

general optimization algorithms
    first-order algorithms 113
    second-order algorithms 113
gradient descent (GD) 41
gradient descent (GD) optimizers 46
    adagrad method 47
    Adaptive Moment Estimation
        (Adam) method 48, 49
    momentum method 46
    RMSprop method 48
gradient descent (GD) variants 43
    application of gradient descent 43-45
graph neural networks (GNNs) 69, 70
graph types 63, 64
    directed graphs 64
    undirected graphs 64
    weighted graphs 65, 66
grid search 100-102

H

hinge loss 54
hyper-parameter tuning or optimization 21

K

Kalman filter 73
    implementation 78-81
Kalman filter algorithm 73
    computation of measurements 75, 76
    filtration of measurements 76, 78
Kalman gain 75
kernel 87
k-nearest neighbors (kNN) 15

L

leaky adagrad 48
linear algebra
    for PCA 30, 31
linear discriminant analysis (LDA) 34-36
linear regression 17
logistic loss 17

M

MachineLearningMastery
    reference link 49
machine learning (ML)
    as mathematical optimization 16
    as predictive tool 21
machine learning (ML) models 41, 102
    Bayesian optimization 103, 104
    grid search 100-102
    random search 97-99
margin 52
Markov Chain Monte Carlo
    Gibbs sampling algorithm 87
    Metropolis-Hastings algorithm 87, 88
mathematical modeling
    digital advertising 26
    energy 25

financial services 25
retail 25
using, as prescriptive tool 24, 25
mathematical optimization 4
problem 4
problem, formulation 5-7
mathematical programming 4
mean squared error (MSE) 43
measurement distribution 75
measurement noise 73
Metropolis-Hastings algorithm 87
examples 89-92
working 88
mini-batch gradient descent 45
ML, as mathematical optimization 16
neural network 18-21
regression 17
ML, as predictive modeling tool 21-23
cybersecurity 24
e-commerce 23
sales and marketing 24
model coefficients 16
momentum method 46

N

network 62
neural network 18-21
noise reduction 38
NoSQL 63

O

operations research
optimization 104-106
optimization use case 66
optimization problem 67
optimization solution 67-69

P

PCA applications
anomaly detection 39
noise reduction 38
polynomial 54
Principal Component Analysis (PCA) 29
applications 36-38
linear algebra 30, 31
principal component (PC) 32
principal subspace 30
probability distribution function (PDF) 74
process noise 73
process variable (PV) 10
programmable logic controller (PLC) 12
proportional-integral-derivative (PID) 10
proportion of variance explained (PVE) 32

R

radial basis function (RBF) 54
random search 97-99
recurrent neural networks (RNNs) 10
regression 17
regularized discriminant analysis 35
RMSprop method 48

S

sampling period 8
second-order algorithms 113
Short Time Fourier Transform (STFT) 8
sigmoid 54
signal 8
signal processing 7
problem 8
problem, formulation 8-10
singular value decomposition (SVD) 33

Index

spectrogram 9
state space model 73
stochastic algorithm 20
stochastic gradient descent
    (SGD) 16, 45, 113
support vector machine (SVM) 51
  implementation 57-59
  kernels 54-56
  support vectors 52-57
support vectors 52
surrogate model 104

T

time shift 9
traveling salesman problem (TSP) 61
tree data structures 62

U

undirected graphs 64

V

vanilla gradient descent 45

W

weighted graphs 65, 66

Packtpub.com

Subscribe to our online digital library for full access to over 7,000 books and videos, as well as industry leading tools to help you plan your personal development and advance your career. For more information, please visit our website.

Why subscribe?

- Spend less time learning and more time coding with practical eBooks and Videos from over 4,000 industry professionals
- Improve your learning with Skill Plans built especially for you
- Get a free eBook or video every month
- Fully searchable for easy access to vital information
- Copy and paste, print, and bookmark content

Did you know that Packt offers eBook versions of every book published, with PDF and ePub files available? You can upgrade to the eBook version at Packtpub.com and as a print book customer, you are entitled to a discount on the eBook copy. Get in touch with us at customercare@packtpub.com for more details.

At www.packtpub.com, you can also read a collection of free technical articles, sign up for a range of free newsletters, and receive exclusive discounts and offers on Packt books and eBooks.

Other Books You May Enjoy

If you enjoyed this book, you may be interested in these other books by Packt:

Machine Learning Security Principles

John Paul Mueller

ISBN: 978-1-80461-885-1

- Explore methods to detect and prevent illegal access to your system
- Implement detection techniques when access does occur
- Employ machine learning techniques to determine motivations
- Mitigate hacker access once security is breached
- Perform statistical measurement and behavior analysis
- Repair damage to your data and applications
- Use ethical data collection methods to reduce security risks

Machine Learning Model Serving Patterns and Best Practices

Md Johirul Islam

ISBN: 978-1-80324-990-2

- Explore specific patterns in model serving that are crucial for every data science professional
- Understand how to serve machine learning models using different techniques
- Discover the various approaches to stateless serving
- Implement advanced techniques for batch and streaming model serving
- Get to grips with the fundamental concepts in continued model evaluation
- Serve machine learning models using a fully managed AWS Sagemaker cloud solution

Packt is searching for authors like you

If you're interested in becoming an author for Packt, please visit `authors.packtpub.com` and apply today. We have worked with thousands of developers and tech professionals, just like you, to help them share their insight with the global tech community. You can make a general application, apply for a specific hot topic that we are recruiting an author for, or submit your own idea.

Share Your Thoughts

Now you've finished *A Handbook of Mathematical Models with Python*, we'd love to hear your thoughts! Scan the QR code below to go straight to the Amazon review page for this book and share your feedback or leave a review on the site that you purchased it from.

`https://packt.link/r/1-804-61670-2`

Your review is important to us and the tech community and will help us make sure we're delivering excellent quality content.

Download a free PDF copy of this book

Thanks for purchasing this book!

Do you like to read on the go but are unable to carry your print books everywhere?

Is your eBook purchase not compatible with the device of your choice?

Don't worry, now with every Packt book you get a DRM-free PDF version of that book at no cost.

Read anywhere, any place, on any device. Search, copy, and paste code from your favorite technical books directly into your application.

The perks don't stop there, you can get exclusive access to discounts, newsletters, and great free content in your inbox daily

Follow these simple steps to get the benefits:

1. Scan the QR code or visit the link below

 `https://packt.link/free-ebook/9781804616703`

2. Submit your proof of purchase
3. That's it! We'll send your free PDF and other benefits to your email directly